# DSM-5-TR
## Desk Reference

*A Concise and Comprehensive Guide to Psychiatric Diagnosis based on the DSM-5-TR*

M. Mastenbjörk M.D.
S. Meloni M.D.

Medical Creations

© **Copyright 2024 by Medical Creations® - All rights reserved.**

This document is geared towards providing exact and reliable information regarding the topic and issue covered. The publication is sold with the idea that the publisher is not required to render accounting, officially permitted, or otherwise qualified services. If advice is necessary, legal or professional, a practiced individual in the profession should be ordered.

From a Declaration of Principles, which was accepted and approved equally by a Committee of the American Bar Association and a Committee of Publishers and Associations.

It is not legal to reproduce, duplicate, or transmit any part of this document electronically or in printed format. Recording this publication is strictly prohibited, and any storage of this document is not permitted unless the publisher gives written permission. All rights reserved.

The information provided herein is truthful and consistent in that any liability, in terms of inattention or otherwise, by any usage or abuse of any policies, processes, or directions contained within is the solitary and utter responsibility of the recipient reader. Under no circumstances will any legal obligation or blame be held against the publisher for any reparation, damages, or monetary loss due to the information herein, either directly or indirectly.

Respective authors own all copyrights not held by the publisher.

The information herein is offered solely for informational purposes and is universal as such. The presentation of the information is without contract or any type of guarantee assurance.

The trademarks that are used are without any consent, and the publication of the trademark is without permission or backing by the trademark owner. All trademarks and brands within this book are for clarifying purposes only and owned by the owners, not affiliated with this document.

# YOU MIGHT ALSO NEED

**DSM-5-TR:**
A Broad Selection of Exercises to Measure
Your Psychiatry Knowledge: Workbook

This QR code is for the Amazon US marketplace, but our workbook is also available in other Amazon marketplaces.

# Contents

Preface .................................................... v

Chapter 1: Neurodevelopmental Disorders ................. 1

Chapter 2: Schizophrenia Spectrum and Other
Psychotic Disorders ..................................... 10

Chapter 3: Bipolar and Related Disorders ................ 17

Chapter 4: Depressive Disorders ......................... 25

Chapter 5: Anxiety Disorders ............................ 31

Chapter 6: Obsessive-Compulsive and Related Disorders ... 38

Chapter 7: Trauma and Stressor-Related Disorders ........ 43

Chapter 8: Dissociative Disorders ....................... 54

Chapter 9: Somatic Symptom and Related Disorders ........ 57

Chapter 10: Feeding and Eating Disorders ................ 62

Chapter 11: Elimination Disorders ....................... 68

Chapter 12: Sleep-Wake Disorders ........................ 70

Chapter 13: Sexual Dysfunctions ......................... 81

Chapter 14: Gender Dysphoria ............................ 89

Chapter 15: Disruptive, Impulse-Control, and Conduct
Disorders .................................................. 92

Chapter 16: Substance-Related and Addictive Disorders ...... 98

Chapter 17: Neurocognitive Disorders ....................... 119

Chapter 18: Personality Disorders .......................... 137

Chapter 19: Paraphilic Disorders ........................... 147

Chapter 20: Cultural Diagnoses and Conditions for
Further Study, Interviewing ................................ 153

References ................................................ 168

# Preface

This *Diagnostic and Statistical Manual of Mental Disorders, Fifth Edition, Text Revision* (DSM-5-TR) desk reference is based on diagnostic criteria in the DSM-5-TR and publications related to psychometrically validated interview tools for establishing diagnoses based on diagnostic criteria in this manual. This desk reference is meant to provide a quick, easy-to-access overview of disorders in the DSM-5-TR. It provides diagnostic criteria for various disorders, organized by the chapter in which they fall in the DSM-5-TR. It also includes coding information for each diagnosis.

This desk reference can be used to find information quickly about a particular diagnosis or help clinicians determine the differences between 2 diagnoses. A clinician may use this desk reference to find the code for a specific diagnosis quickly or to review the diagnostic criteria to assist with making a diagnosis.

For more comprehensive information, a clinician should use the full DSM-5-TR manual. For instance, DSM-5-TR provides detailed information about diagnostic features, specifiers, associated features, and prevalence rates of mental health disorders. The desk reference does not include such information because it is a simple, easy-to-use guide. Nonetheless, it is a portable tool, quickly providing clinicians with information about DSM-5-TR diagnoses.

This manual can also be used to study for a DSM-5-TR examination or to reference a code or diagnostic criteria for quick confirmation. Clinicians should use the entire manual, when necessary, to obtain additional information for making a differential diagnosis, such as prevalence or diagnostic features. In some cases, such as when details about a specifier are needed, this desk reference may not be sufficient to make a complete, accurate diagnosis with proper coding and specifiers.

# CHAPTER 1

# Neurodevelopmental Disorders

The DSM-5-TR neurodevelopmental disorders are a category of disorders with onset in the early developmental period, typically before a child enters school. These disorders involve deficits or differences in brain functioning that result in difficulties functioning in personal, social, academic, or occupational contexts. There is a high rate of co-occurrence among the different neurodevelopmental disorders; for instance, ADHD may occur alongside specific learning disorders.

A person must show symptoms of a disorder and an impairment in functioning to be diagnosed with a neurodevelopmental disorder. The specific disorders included in this chapter are detailed below.

## Intellectual Developmental Disorder

This disorder, also called an intellectual disability, involves intelligence and adaptive skills deficits. Diagnosis requires a person to meet the following criteria:

- Deficits in intellectual functions such as reasoning, problem-solving, planning, thinking, judgment, and learning
- Deficits in adaptive functioning skills that result in difficulty meeting the demands of daily life
- Symptom onset during the developmental period

Specifiers include severity level, which can be mild, moderate, severe, or profound.

Coding is as follows:

- Mild (F70)
- Moderate (F71)
- Severe (F72)
- Profound (F73).

Related disorders include:

Global Developmental Delay (F88), which is diagnosed when:

- Children under 5 years old fail to meet typical developmental milestones related to intelligence
- Unable to complete systematic assessments of intellectual functioning.

Unspecific Intellectual Developmental Disorder (F79) is a diagnosis made in exceptional circumstances when:

- Children over the age of 5 show signs of intellectual deficits
- Assessment is difficult or impossible due to sensory or physical impairments like blindness or deafness, comorbid mental disorders, or severely problematic behaviors.

## Communication Disorders

This category of disorders involves impairments in speech, language, and communication. Disorders in this category include:

Language Disorder (F80.2), which is diagnosed in children who have:

Difficulties related to the acquisition and use of language, resulting from deficits in comprehension or production, which can include the following:

- Reduced vocabulary
- Impaired ability to form grammatically correct sentences
- Impaired discourse abilities, meaning the child struggles to use vocabulary and connect sentences when having conversations or describing topics

Speech Sound Disorder (F80.0), which is characterized by:
- Ongoing difficulty with the production of speech sounds, which makes speech unintelligible or prevents a person from communicating verbally
- Symptoms also limit the ability to communicate in social settings, school, or the workplace.

Childhood-Onset Fluency Disorder (F80.81), also called stuttering, which results in
- Disruptions in the typical fluency and time patterning of speech, which last over time and cause sound/syllable repetitions, sound prolongations, pauses within a word, pauses in speech, word substitutions, words produced with excess physical tension, and monosyllabic whole-word repetitions
- The speech disturbance must cause anxiety about speaking or impairments in communication, social interaction, school performance, or work performance.

Social (Pragmatic) Communication Disorder (F80.82), which involves:
- Difficulties using both verbal and non-verbal communication in social settings, as evidenced by symptoms including deficits in behaviors like:
  - Greeting and information sharing during social interactions
  - Changing communication strategies to align with context
  - Following rules for conversation and storytelling
  - Making inferences.

Unspecified Communication Disorder (F80.9), a diagnosis used when:
- A person has symptoms representative of a communication disorder
- Symptoms cause clinically significant distress or impairment but do not meet the full criteria for another communication disorder or neurodevelopmental disorder.

## Autism Spectrum Disorder

This disorder (F84.0) involves:

- Ongoing impairments in social communication and interactions with restrictive and repetitive behavior patterns, interests, or activities.
- Deficits in social communication and interactions in autism spectrum disorder manifest by all of the following symptoms:
    - Impairment in social-emotional reciprocity, such as difficulty with back-and-forth conversation or inability to respond to social interactions
    - Difficulty using nonverbal communication during social interactions, as evidenced by deficits like difficulty with eye contact or understanding the use of gestures
    - Struggling to develop, maintain, or understand social relationships, which may range from difficulty adjusting one's behavior to different environments to a total lack of interest in friendships
- Restricted, repetitive patterns of behavior are present, as indicated by 2 or more of the following:
    - Stereotypical or repetitive motor behavior, object use, or speech patterns, such as lining up toys or using idiosyncratic phrases
    - Insisting that things always be the same and that routines be maintained
    - Fixated interests that are unusual in their intensity or focus
    - Unusual reactivity to sensory stimuli, which may include either over or under-reactivity

Additional diagnostic criteria:

- Symptoms must be present in the early developmental period
- Symptoms must cause clinically significant impairment
- Symptoms cannot be explained by intellectual development disorder (intellectual disability) or global developmental delay.

Specifiers include:

- Requiring very substantial support, requiring substantial support, or requiring support

- With or without accompanying intellectual impairment
- With or without accompanying language impairment
- Associated with a known genetic or other medical condition or environmental factor (requires the use of additional code to identify the associated genetic or other medical condition)
- Associated with a neurodevelopmental, mental, or behavioral problem
- With catatonia (uses additional code F06.1)

## Attention-Deficit/Hyperactivity Disorder (ADHD)

ADHD involves an ongoing pattern of inattention, hyperactivity/impulsivity, or both. For diagnosis, a child must show:

- At least 6 symptoms of inattention, hyperactivity/impulsivity, or both.
- Those aged 17 and older may be diagnosed when they show 5 or more symptoms.
- Symptoms of inattention include:
  - Not paying attention to details
  - Making careless mistakes when performing schoolwork or job-related tasks
  - Struggling to maintain attention to tasks or play activities
  - Appearing not to listen when being spoken to directly
  - Tending not to follow through with instructions and failing to finish schoolwork, job tasks, or chores
  - Struggling to organize tasks and activities, which can manifest as poor time management or messy work
  - Avoidance or dislike of tasks like schoolwork or lengthy reports because they require sustained mental effort
  - Tendency to lose important items like a wallet, phone, or keys
  - Being easily distracted by outside stimuli
  - Tending to be forgetful in daily activities such as doing chores or returning calls
- Symptoms of hyperactivity and impulsivity include:
  - Frequent fidgeting
  - Leaving one's seat when remaining seated is expected
  - Running or climbing when it's not appropriate to do so

- Inability to play or engage in leisure activities quietly
- Being constantly on the go
- Tending to talk excessively
- Blurting out answers before a question is completed
- Struggling to wait one's turn
- Interrupting others or intruding on their conversations or activities

Additional diagnostic criteria:

- Symptoms must persist for at least 6 months
- Several symptoms must have been present before age 12.
- Symptoms must also occur in at least 2 settings, such as at home and school.

Specifiers include:

- Combined presentation (F90.2)- Criterion for both inattention and hyperactivity/impulsivity are met
- Predominantly inattentive presentation (F90.0)- Only criteria for inattention are met
- Predominantly hyperactive/impulsive presentation (F90.1)- Only criteria for hyperactivity/impulsivity are met
- In partial remission- When full criteria were met previously but have not been met for the last 6 months, yet symptoms still cause impairment
- Mild, moderate, or severe

Related disorders include:

Other Specified Attention-Deficit/Hyperactivity Disorder (F90.8), which is diagnosed when a person has:

- Symptoms of attention-deficit/ hyperactivity disorder
- Symptoms cause significant distress or impairment despite not meeting full diagnostic criteria
- The clinician specifies the reason full diagnostic criteria are not met

Unspecified Attention-Deficit/Hyperactivity Disorder (F90.9), which is diagnosed when a person has:

- Symptoms of attention-deficit/ hyperactivity disorder
- Symptoms cause significant distress or impairment despite not meeting full diagnostic criteria
- The clinician does not specify the reason full criteria are not met

## Specific Learning Disorder

Specific Learning Disorder is diagnosed in children who experience challenges with learning and academic skills. For diagnosis:

- A child must experience one or more of the following symptoms for at least 6 months:
  - Showing slow or effortful reading skills
  - Struggling to understand the meaning of what one has read
  - Struggling with spelling
  - Having a difficult time with written expression as evidenced by deficits like grammatical errors or poor paragraph organization
  - Struggling to master number sense and calculation
  - Finding mathematical reasoning to be challenging, such that one has severe difficulty solving math problems

Additional diagnostic criteria:

- Affected academic skills must be significantly below what is expected for an individual's age
- Standardized achievement tests and clinical assessment must substantiate deficits.
- The learning difficulties begin during the school years.

Specifiers include:

- With impairment in reading (F81.0): word reading accuracy, reading rate or fluency, reading comprehension
- With impairment in written expression (F81.81): spelling accuracy, grammar and punctuation accuracy, clarity or organization of written expression

- With impairment in mathematics (F81.2): number sense, memorization of arithmetic facts, accurate or fluent calculation, accurate math reasoning
- Mild, moderate, or severe

## Motor Disorders

Motor disorders are also included among the neurodevelopmental disorders. Disorders in this category include:

Developmental Coordination Disorder (F82), which involves

- Coordinated motor skills that are significantly below what is expected given a person's age and result in behaviors like clumsiness and slow performance of motor movements that interfere with daily functioning.
- Onset is in the early developmental period.

Stereotypic Movement Disorder (F98.4), which is characterized by

- Repetitive motor behaviors like hand shaking or body rocking, which serve no purpose and interfere with daily functioning

Specifiers:

- With or without self-injurious behavior
- Associated with a known genetic or other medical condition, neurodevelopmental disorder, or environmental factor
- Mild, moderate, or severe

Tourette's Disorder (F95.2), which is characterized by:

- Multiple motor tics and at least one verbal tic, which may vary in frequency but have lasted over a year
- Onset is before age 11.

Persistent (Chronic) Motor or Vocal Tic Disorder (F95.1), which involves:

- Single or multiple motor or verbal tics but not both
- The frequency of tics can vary, but they have lasted over a year.
- Requires a specifier of vocal tics only or motor tics only

Provisional Tic Disorder (F95.0), which is diagnosed when a person has

- Single or multiple motor or verbal tics that have persisted for less than a year.
- The criteria for Tourette's or persistent (chronic) motor or vocal tic disorder have not been met

Related disorders include:

Other Specified Tic Disorder (F95.8), diagnosed when a person has:

- Symptoms of a tic disorder that do not meet full diagnostic criteria for another tic disorder or neurodevelopmental disorder or cause clinically significant impairment in one or more areas of functioning.
- The clinician must specify why the presentation doesn't meet the full criteria.

Unspecified Tic Disorder (F95.9), diagnosed when a person has:

- Symptoms of a tic disorder that do not meet full diagnostic criteria but impair the ability to function
- The clinician does not specify the reason the presentation doesn't meet the full criteria.

Other Specified Neurodevelopmental Disorder (F88) diagnosed when a person has:

- Symptoms of a neurodevelopmental disorder that do not meet full diagnostic criteria for another neurodevelopmental disorder
- The clinician specifies the reason full criteria are not met, including the specific reason the full diagnostic criteria are not met, with the specific reason. An example is neurodevelopmental disorder associated with prenatal alcohol exposure.

Unspecified Neurodevelopmental Disorder (F89), diagnosed when a person has:

- Symptoms of a neurodevelopmental disorder that do not meet full diagnostic criteria for another neurodevelopmental disorder
- The clinician does not specify the reason full criteria are not met.

## CHAPTER 2

# Schizophrenia Spectrum and Other Psychotic Disorders

The schizophrenia spectrum and other psychotic disorders chapter in the DSM-5-TR includes schizophrenia, other psychotic disorders, and schizotypal personality disorder. Disorders in this chapter involve delusions, persistent beliefs maintained despite evidence that the belief is untrue. These disorders also involve hallucinations, which are perceptions a person experiences without external stimuli. Additional symptoms of the disorders in this chapter include disorganized thinking and speech, disorganized or abnormal behavior, and negative symptoms, which include lack of emotional expression, lack of motivation (avolition), diminished speech (alogia), lack of pleasure (anhedonia), and asociality.

The specific disorders included in this chapter are detailed below.

## Schizotypal Personality Disorder

Simply listed at the start of this chapter, the full criteria for schizotypal personality disorder are found in the "Personality Disorders" chapter.

## Delusional Disorder

Delusional disorder (F22) involves:

- Delusions lasting at least one month
- The person cannot have ever met the criteria for schizophrenia if they are diagnosed with this disorder

- Hallucinations may be present but are related to the delusional theme.
- Impaired functioning is minimal.
- The delusion must last for a month or more.

Specifiers include:

- Type of delusion:
  - Erotomanic: Belief that someone else is in love with the person
  - Grandiose: Belief that a person has a wonderful but unrecognized talent or has made an essential discovery
  - Jealous: Belief that one's spouse or significant other is unfaithful
  - Persecutory: Belief that one is being conspired against, poisoned, followed, harassed, etc.
  - Somatic: Beliefs related to bodily functions or sensations
  - Mixed: Utilized when there is no single delusion type that predominates
  - Unspecified: Utilized when the predominant delusion type is unclear or nonspecific
  - With bizarre content
- First episode, currently in an acute episode
- First episode, currently in partial remission
- First episode, currently in full remission
- Multiple episodes, currently in acute episode
- Multiple episodes, currently in partial remission
- Multiple episodes, currently in full remission
- Continuous
- Unspecified
- Severity of primary symptoms of psychosis (specifier not required)

## Brief Psychotic Disorder

This disorder (F23) is diagnosed in people who show at least one of the following symptoms:

1. Delusions
2. Hallucinations
3. Disorganized speech
4. Disorganized/catatonic behavior

- To be diagnosed, at least 1, 2, or 3 MUST be present
- Symptoms must last at least one day but under a month, with the person returning to their typical level of functioning after symptoms resolve.

Specifiers include:

- With marked stressors
- Without marked stressors
- With peripartum onset
- With catatonia (uses additional code, F06.1)
- Severity of primary symptoms of psychosis (specifier not required)

## Schizophreniform Disorder

This disorder (F20.81) requires a person to show the following 2 or more of the following symptoms:

1. Delusions
2. Hallucinations
3. Disorganized speech
4. Disorganized/catatonic behavior
5. Negative symptoms, like lack of motivation or emotional expression

- At least one symptom must be 1, 2, or 3
- The symptoms must be present for at least one month but under 6 months
- A provisional diagnosis can be made if waiting to determine if a person recovers within 6 months is unreasonable.

Specifiers include:

- With good prognostic features
- Without good prognostic features
- With catatonia (uses additional code, F06.1)
- Severity of primary symptoms of psychosis (specifier not required)

## Schizophrenia

This disorder (F20.9) requires a person to show 2 or more of the following symptoms:

1. Delusions
2. Hallucinations
3. Disorganized speech
4. Disorganized/catatonic behavior
5. Negative symptoms, like lack of motivation or emotional expression
   - At least one symptom must be 1, 2, or 3.
   - To be diagnosed, the symptoms above must be present for at least one month, but signs of psychological disturbance must last at least 6 months.
   - During the 6-month period, a person must show the symptoms above for at least one month, but they can also experience residual or prodromal periods during which they show only negative symptoms or weakened features associated with symptoms 1, 2, or 3 above.

Specifiers include:
- First episode, currently in an acute episode
- First episode, currently in partial remission
- First episode, currently in full remission
- Multiple episodes, currently in acute episode
- Multiple episodes, currently in partial remission
- Multiple episodes, currently in full remission
- Continuous
- Unspecified
- With catatonia (uses additional code, F06.1)
- Severity of primary symptoms of psychosis (specifier not required)

## Schizoaffective Disorder

Schizoaffective Disorder involves:

- Symptoms of a major mood episode, which can be depressive or manic, co-occurring with symptoms of schizophrenia
- At some point during the lifetime of the illness, experiencing delusions or hallucinations for at least 2 weeks without any major mood episode symptoms
- Major mood episode symptoms are present for most of the illness

Specifiers include:

- Bipolar type (F25.0), diagnosed in those with manic episodes; depressive episodes may also be present during the illness
- Depressive type (F25.1), diagnosed in those who only experience depressive episodes during the illness
- With catatonia (use additional code, F06.1)
- First episode, currently in an acute episode
- First episode, currently in partial remission
- First episode, currently in full remission
- Multiple episodes, currently in acute episode
- Multiple episodes, currently in partial remission
- Multiple episodes, currently in full remission
- Unspecified
- Severity of primary symptoms of psychosis (specifier not required)

## Other psychotic disorders

Other psychotic disorders in this chapter are as follows:

Substance/Medication-Induced Psychotic Disorder, which is characterized by experiencing:

- Hallucinations or delusions, with evidence that these symptoms developed during or soon after a substance or medication intoxication, use, or withdrawal

Coding depends upon the specific substance or medication involved and whether a person has a substance use disorder.

Psychotic Disorder Due to Another Medical Condition is diagnosed when a person has either hallucinations or delusions resulting directly from another medical condition.

Coded as:

- F06.2 if the primary symptom is delusions
- F06.0 if the primary symptom is hallucinations.
- You do not have to specify the severity.
- You can include the medical diagnosis in the name of the psychiatric disorder but should code the medical condition separately. List and code the medical condition immediately before listing the mental disorder.

Catatonia Associated with Another Mental Disorder, also called the catatonia specifier (F06.1), which is diagnosed when a person has:

- 3 or more of the following:
  - Stupor
  - Catalepsy
  - Waxy flexibility
  - Mutism
  - Negativism
  - Posturing
  - Mannerism
  - Stereotypy
  - Agitation
  - Grimacing
  - Echolalia
  - Echopraxia.
- You should indicate the associated mental disorder when listing the name of the condition. Code the mental diagnosis first before coding the catatonia with specifier.

Catatonic Disorder Due to Another Medical Condition (F06.1), diagnosed when a person has:

- At least 3 of the catatonic symptoms noted above in catatonia associated with another medical condition
- The catatonic symptoms are caused directly by the medical problem.

Unspecified Catatonia, diagnosed when a person has the following:
- Symptoms of catatonia
- The cause is unclear, or full criteria are not met

Coded as:
- R29.818, followed by F06.1 for unspecified catatonia.

Other Specified Schizophrenia Spectrum and Other Psychotic Disorder (F28), diagnosed when a person has:

- Symptoms characteristic of a disorder in this chapter, but full criteria are not met
- The clinician specifies the reason full criteria are not met.

Unspecified Schizophrenia Spectrum and Other Psychotic Disorder (F29), diagnosed when a person has:

- Symptoms characteristic of a disorder in this chapter, but full criteria are not met
- The clinician does not specify the reason full criteria are not met.

# CHAPTER 3

# Bipolar and Related Disorders

The bipolar and related disorders chapter in the DSM-5-TR includes disorders that are seen as a bridge between the schizophrenia spectrum and other psychotic disorders and depressive disorders. Primary disorders in this chapter include bipolar I disorder, bipolar II disorder, and cyclothymic disorder. These disorders are characterized by mood episodes involving mania or depression.

The specific disorders in this chapter are detailed below.

## Bipolar I Disorder

To be diagnosed with bipolar I disorder, a person must have at least one manic episode; they may also experience hypomanic or major depressive episodes.

Criteria for a manic episode are as follows:
- Experiencing a distinct period during which one's mood is elevated, expansive, and irritable, and one experiences a persistent increase in energy and activity; this period must last at least one week or for any duration if a person must be hospitalized.
- Showing at least 3 of the following symptoms during the mood episode:
  - High self-esteem or grandiosity
  - Reduced need for sleep
  - Increase in talking or pressure to keep talking
  - Showing a flight of ideas or experiencing, subjectively, that one's thoughts are racing
  - Being easily distracted

- ○ An increase in goal-directed activity, which can be in social settings, at work, or school; alternatively, the display of psychomotor agitation
- ○ Engaging in activities that can cause serious consequences, such as excessive shopping, unprotected sex, or unwise business investments
- Evidence that the mood episode is significant enough to cause marked dysfunction or to necessitate hospitalization to prevent harm to self or others; alternatively, the presence of psychosis.
- The episode does not occur as a result of a physiological response to drugs of abuse, medications, or other medical conditions.

If a person experiences a hypomanic episode in the context of bipolar I disorder, they must meet the following criteria:

- Experiencing an elevated, expansive, or irritable mood and an increase in energy or activity lasting at least 4 days
- Showing at least 3 of the following symptoms during the mood episode:
  - ○ High self-esteem or grandiosity
  - ○ Reduced need for sleep
  - ○ Increase in talking or pressure to keep talking
  - ○ Showing a flight of ideas or experiencing, subjectively, that one's thoughts are racing
  - ○ Being easily distracted
  - ○ An increase in goal-directed activity, which can be in social settings, at work, or school; alternatively, the display of psychomotor agitation
- Engaging in activities that can cause serious consequences, such as excessive shopping, unprotected sex, or unwise business investments
- Experiencing a clear change in functioning that is not only uncharacteristic for the person but also observable by others
- Absence of marked impairment or need for hospitalization; also, absence of psychotic features

If a person experiences a depressive episode in the context of bipolar I disorder, they must meet the following criteria:

- Experiencing 5 or more of the following symptoms during a 2-week period, with at least one symptom being depressed mood or loss of interest or pleasure with usual activities.
    - A depressed mood nearly every day for most of the day. A depressed mood can involve feeling sad, empty, or hopeless or observation by others that one appears sad.
    - Reduced interest or pleasure with all or nearly all usual activities
    - Significant change in weight or appetite, which can include either increases or decreases in appetite or weight
    - Insomnia or increase in sleep
    - Experiencing either psychomotor retardation or agitation nearly daily, which is observable by others
    - Sense of fatigue or low energy
    - Feeling worthless or inappropriately guilty
    - Difficulty with thinking, concentrating, or making decisions
    - Recurring thoughts of death, suicidal ideation without a plan, or suicide attempts or plans

Coding is based on severity, most recent type of episode, and the features of the disorder, which are specific as follows:

- Mild, with the current or most recent episode being manic: F31.11
- Moderate, with the current or most recent episode being manic: F31.12
- Severe, with the current or most recent episode being manic: F31.13
- Mild, with the current or most recent episode being depressed: F31.31
- Moderate, with the current or most recent episode being depressed: F31.32
- Severe, with the current or most recent episode being depressed: F31.4
- With psychotic features, with the current or most recent episode being manic: F31.2 (coded this way regardless of severity if psychotic features present)
- With psychotic features, with the current or most recent episode being depressed: F31.5 (coded this way irrespective of severity if psychotic features present)

- In partial remission, with the current or most recent episode being manic: F31.73
- In partial remission, with the current or most recent episode being depressed: F31.75
- In full remission, with the current or most recent episode being manic: F31.74
- In full remission, with the current or most recent episode being depressed: F31.76
- Unspecified, with the current or most recent episode being manic: F31.9
- Unspecified, with the current or most recent episode being hypomanic: F31.9
- Unspecified, with the current or most recent episode being depressed: F31.9

Specifiers include:

- With anxious distress
- With mixed features
- With rapid cycling
- With melancholic features
- With atypical features
- With mood-congruent psychotic features
- With mood-incongruent psychotic features
- With catatonia (uses additional code, F06.1)
- With peripartum onset
- With seasonal pattern

## Bipolar II Disorder

To be diagnosed with bipolar II disorder (F31.81), a person must experience:

- At least one hypomanic episode and at least one major depressive episode
- A person cannot be diagnosed with bipolar II disorder if they have ever experienced a manic episode.

- When a person experiences a hypomanic episode in the context of bipolar II disorder, they must meet the following criteria:
  - Experiencing an elevated, expansive, or irritable mood and an increase in energy or activity lasting at least 4 consecutive days
  - Showing at least 3 of the following symptoms during the mood episode:
    - High self-esteem or grandiosity
    - Reduced need for sleep
    - Increase in talking or pressure to keep talking
    - Showing a flight of ideas or experiencing, subjectively, that one's thoughts are racing
    - Being easily distracted
    - An increase in goal-directed activity, which can be in social settings, at work, or school; alternatively, the display of psychomotor agitation
- Engaging in activities that can cause serious consequences, such as excessive shopping, unprotected sex, or unwise business investments
- Experiencing a clear change in functioning that is not only uncharacteristic for the person but also observable by others
- Absence of marked impairment or need for hospitalization; also, absence of psychotic features

When a person experiences a depressive episode in the context of bipolar II disorder, they must meet the following criteria:

- Experiencing 5 or more of the following symptoms during 2 consecutive weeks with at least one symptom being depressed mood or loss of interest or pleasure with usual activities
  - A depressed mood, nearly every day for most of the day, can involve feeling sad, empty, or hopeless or observation by others that one appears sad
  - Reduced interest or pleasure with all or nearly all usual activities
  - Significant change in weight or appetite, which can include either increases or decreases in appetite or weight
  - Insomnia or increase in sleep

- Experiencing either psychomotor retardation or agitation nearly daily, which is observable by others
- Sense of fatigue or low energy
- Feeling worthless or inappropriately guilty
- Difficulty with thinking, concentrating, or making decisions
- Recurring thoughts of death, suicidal ideation without a plan, or suicide attempts or plans

Specifiers include:

- Whether the current or most recent episode is hypomanic or depressed
  - If hypomanic, whether the episode is:
    - With anxious distress
    - With mixed features
    - With rapid cycling
    - With peripartum onset
    - With seasonal pattern
  - If depressed, whether the episode is:
    - With anxious distress
    - With mixed features
    - With rapid cycling
    - With melancholic features
    - With atypical features
    - With mood-congruent psychotic features
    - With mood-incongruent psychotic features
    - With catatonia
    - With peripartum onset
    - With seasonal pattern
    - In partial remission
    - In full remission
- Severity of mild, moderate, or severe

## Cyclothymic Disorder

Cyclothymic disorder (F34.0) is diagnosed in people who have shown:

- Numerous periods of hypomanic symptoms that don't meet full criteria for a hypomanic episode, combined with multiple periods with

- depressive symptoms that do not meet full criteria for a hypomanic episode
- The disturbance lasts at least 2 years in adults and at least 1 year in children and adolescents.

Additional criteria for cyclothymic disorder include:

- Presence of symptoms for at least half the time during the 2-year disturbance, with no more than 2 months at a time without the presence of symptoms
- Never meeting full criteria for a depressive, manic, or hypomanic episode

Specifiers:

- With anxious distress

## Other Bipolar and Related Disorders

Other bipolar and related disorders in this chapter are as follows:

Substance/Medication-Induced Bipolar and Related Disorder, which is diagnosed in people who have:

- A mood disturbance involving elevated, expansive, or irritable mood and increased activity or energy
- Occurs during or soon after a substance or medication intoxication, withdrawal, or exposure

Coding is dependent upon the specific substance and whether there is a use disorder

Specifiers include:

- With onset during intoxication
- With onset during withdrawal
- With onset after medication use

Bipolar and Related Disorder Due to Another Medical Condition, which is diagnosed in people who have a mood disturbance involving elevated, expansive, or irritable mood and increased activity or energy that occurs as a direct result of a medical condition

Specified as:

- With manic features (F06.33)
- With a manic-or hypomanic-like episode (F06.33)
- With mixed features (F06.34)

Other Specified Bipolar and Related Disorder (F31.89), which is diagnosed when people have:

- Symptoms representing a bipolar and related disorder, that do not meet the full criteria for any disorder in this chapter
- The clinician communicates why the presentation does not meet criteria for another disorder

Unspecified Bipolar and Related Disorder (F31.9), which is diagnosed when people have:

- Symptoms representing a bipolar and related disorder, but do not meet full criteria for any disorder in this chapter.
- The clinician does not communicate why the presentation does not meet full criteria for another disorder.

Unspecified Mood Disorder (F39), which is diagnosed when a person has:

- Symptoms characteristic of a mood disorder that do not meet the full criteria for any disorders in the bipolar or depressive disorders diagnostic classes
- The clinician cannot decide between an unspecified bipolar and related disorder and an unspecified depressive disorder.

# CHAPTER 4

# Depressive Disorders

The depressive disorders chapter in the DSM-5-TR includes disorders that have the presence of a sad, empty, or irritable mood in common. Other features of these disorders include other changes, such as somatic or cognitive complaints, that significantly impair an individual's functioning.

The specific disorders in this chapter are detailed below.

## Disruptive Mood Dysregulation Disorder

Disruptive mood dysregulation disorder (F34.81) is diagnosed in children between the ages of 6 years and 18 months. Symptoms must be present before the age of 10 years.

Diagnostic criteria include:

- Severe and repeated temper outbursts that can consist of verbal or behavioral aggression and are out of proportion to the situation provoking them
- Temper outbursts inconsistent with one's developmental level
- Temper outbursts occurring an average of at least 3 times per week
- Persistently irritable or angry mood between temper outbursts, which is observable by other people
- A child must show criteria for at least 12 months, with no period of more than 3 consecutive months, with all criteria met
- The criteria must be met in at least 2 of 3 settings (home, school, and with peers).

## Major Depressive Disorder

Major depressive disorder is diagnosed in people who show:

- At least 5 of the following symptoms over a period lasting at least 2 weeks, with at least one of the symptoms being depressed mood or loss of pleasure:
  - Depressed mood most days, throughout the day, as indicated by either a subjective report of feeling sad, hopeless, and empty or by an observation made by others that one is tearful, etc.
  - Reduced interest and pleasure with usual activities
  - Change in weight, which can occur as either weight loss or weight gain or increase or decrease in appetite
  - Either insomnia or sleeping too much
  - Psychomotor agitation or retardation that others can observe
  - Experience fatigue or energy loss nearly daily
  - Feeling worthless or inappropriately guilty
  - Difficulty with thinking, concentrating, or making decisions
  - Recurring thoughts of death, suicidal ideation without a plan, or suicide attempts or plans

Coding is based on the severity and whether a person experiences a single or recurrent episode, as follows:

- Mild, single episode: F32.0
- Mild, recurrent episode: F33.0
- Moderate, single episode: F32.1
- Moderate, recurrent episode: F33.1
- Severe, single episode: F32.2
- Severe, recurrent episode: F33. 3
- With psychotic features, single episode: F32.3 (coded this way regardless of severity if psychotic features are present)
- With psychotic features, recurrent episode: F33.3 (coded this way irrespective of severity if psychotic features are present)
- In partial remission, single episode: F32.4
- In partial remission, recurrent episode: F33.41
- In full remission, single episode: F32.5

- In full remission, recurrent episode: F33.42
- Unspecified, single episode: F32.9
- Unspecified, recurrent episode: F33.9

Specifiers include:
- With anxious distress
- With mixed features
- With melancholic features
- With atypical features
- With mood-congruent psychotic features
- With mood-incongruent psychotic features
- With catatonia
- With peripartum onset
- With seasonal pattern

## Persistent Depressive Disorder

Persistent depressive disorder (F34.1) is diagnosed in people who have:

- A depressed mood nearly all day, most days, for at least 2 years (or one year with irritable mood meeting diagnostic criteria for children and adolescents).
- During the period of depression, a person must show at least 2 of the following symptoms to meet diagnostic criteria:
  - Lack of appetite or overeating
  - Insomnia or excessive sleep
  - Lack of energy or fatigue
  - Low self-esteem
  - Difficulty concentrating or making decisions
  - Feeling helpless
- To be diagnosed, a person cannot be without symptoms for more than 2 consecutive months.

Specifiers include:

- With anxious distress
- With atypical features
- In partial remission
- In full remission
- Early onset (before age of 21)
- Late-onset (at age 21 or older)
- With pure dysthymic syndrome
- With persistent major depressive disorder
- With intermittent major depressive episodes, with current episode
- With intermittent major depressive episodes, without current episode
- Severity of mild, moderate, or severe

## Premenstrual Dysphoric Disorder

Premenstrual dysphoric disorder (F32.81) is diagnosed in people who show

- At least 5 symptoms that present during the final week before the onset of menses for the majority of menstrual cycles.
- The symptoms begin to show improvement within a few days of the start of menses and become minimal or absent the week after menses.
- To be diagnosed, a person must show at least one of the following symptoms:
  - Noticeable mood instability, which can be evidenced by mood swings, becoming suddenly tearful, or showing increased rejection sensitivity
  - Noticeable anger or irritability or increases in conflict with others
  - Noticeably depressed mood, hopeless feelings, or self-deprecating thoughts
  - Noticeable anxiety, tension, or feelings that one is keyed up or on edge
- They must also show at least one of the following additional symptoms, totaling 5 symptoms between the 2 categories:
  - Reduced interest in typical activities
  - Difficulty concentrating
  - Feelings of lethargy, fatigue, or noticeable lack of energy

- Significant change in appetite, which can include overeating or cravings for specific foods
- Insomnia or excessive sleeping
- Feeling overwhelmed or out of control
- Physical symptoms like swollen or tender breasts, muscle and joint pain, bloating, or weight gain

## Other Depressive Disorders

Other depressive disorders in this chapter are as follows:

Substance/medication-induced depressive disorder, which is diagnosed when a person has:

- A depressed mood or reduced interest in or pleasure with usual activities
- There is evidence that the disturbance appeared during or soon after intoxication, withdrawal, or exposure involving a medication or substance

Coding depends upon the type of substance or medication and whether a person has a use disorder.

Depressive disorder due to another medical condition, which is diagnosed when a person has:

- A depressed mood or reduced interest in or pleasure with usual activities
- There is evidence that the disturbance was directly caused by a medical condition specified as with depressive features (F06.31), with major depressive-like episode (F06.32), or with mixed features (F06.34).

Other specified depressive disorder (F32.89), which is diagnosed when a person has:

- Symptoms of a depressive disorder that do not meet the criteria for one of the other disorders in this chapter
- The clinician specifies the reason full criteria for another depressive disorder are not met.

Unspecified depressive disorder (F32.A), which is diagnosed when a person has:

- Symptoms of a depressive disorder that do not meet the criteria for one of the other disorders in this chapter.
- The clinician does not specify the reason full criteria for another depressive disorder are not met.

Unspecified mood disorder (F39), which is diagnosed when a person has:

- Symptoms characteristic of a mood disorder but does not meet the full criteria for any disorders in the bipolar or depressive disorders diagnostic classes
- The clinician cannot decide between an unspecified bipolar and related disorder and an unspecified depressive disorder.

## CHAPTER 5

# Anxiety Disorders

The anxiety disorders chapter in the DSM-5-TR includes disorders that involve excessive fear and worry coupled with behavioral responses. There are various anxiety disorders, and they differ in the situation or object that results in a fear response. Anxiety disorders differ from typical, day-to-day fear or response to stressors, as they are persistent, often causing anxiety lasting at least 6 months.

The specific disorders in this chapter are detailed below.

## Separation Anxiety Disorder

Separation anxiety disorder (F93.0) involves excessive fear or anxiety that rises to the level of being developmentally inappropriate when separating from an attachment figure.

To be diagnosed, a person must meet the following criteria:

- Recurring extreme distress when anticipating separation from or separating from home or an attachment figure
- Worrying excessively about losing significant attachment figures or worrying that attachment figures will suffer harm through illness, injury, disasters, or death
- Worrying excessively that an adverse event like kidnapping or an accident will occur, resulting in separation from an attachment figure
- Being reluctant or refusing to go out or away from home because of fears related to separation
- Excessive fear about being alone or without a major attachment figure, whether at home or another setting

- Ongoing reluctance or refusal to sleep at a place away from home or to go to sleep without the presence of the attachment figure
- Nightmares related to separation
- The experience of physical symptoms, like headaches, stomachaches, nausea, and vomiting when separated from the attachment figure or when separation is anticipated
- The fear related to separation must last at least 4 weeks in children and adolescents and 6 months or more in adults.

## Selective Mutism

Selective mutism (F94.0) is diagnosed in those who repeatedly fail to speak in social situations when there is an expectation of speaking despite being able to speak in other settings. Other criteria include:

- Disturbance lasting at least one month, which is not the first month of school
- Disturbance interferes with achievement at work or school or with social communication
- Disturbance is not explained by lack of knowledge or comfort with the spoken language
- Disturbance does not occur exclusively in the context of autism, schizophrenia, or another psychotic disorder and is not better explained by a communication disorder

## Specific Phobia

A specific phobia is diagnosed in individuals who have:

- Noticeable fear or anxiety related to a specific object or situation, such as flying, heights, animals, seeing blood, or getting an injection.
- In children, fear and anxiety may present in the form of crying, tantrums, freezing, or clinging

Other criteria for diagnosis include:

- The situation or object triggering fear nearly always results in immediate fear or anxiety.

- The individual avoids the situation or object that triggers fear or anxiety or endures it with intense fear or anxiety.
- The fear or anxiety related to the phobia exceeds the actual danger presented by the object or situation.
- The fear or anxiety is persistent, as evidenced by lasting for at least 6 months.

Coding is based on the source of the phobia, as follows:

- Animal: F40.218
- Natural environment: F40.228
- Blood-injection injury: F40.23x (F40.230-fear of blood; F40.231-fear of injections and transfusions; F40.232-fear of other medical care; F40.233-fear of injury)
- Situational: F40.248 (includes airplanes, elevators, and enclosed spaces)
- Other: F40.298 (includes situations that may lead to choking/vomiting, loud sounds, or costumed characters)

## Social Anxiety Disorder

Social anxiety disorder (F40.10) is diagnosed in people who have:

- Noticeable fear or anxiety related to social situations that may result in scrutiny by others. People with social anxiety disorder may be fearful of situations, including having conversations, meeting new people, being observed, or giving a speech.
- Fear that one will show anxiety symptoms or act in a way that results in negative evaluation by others
- Social situations almost always result in fear or anxiety.
- The individual avoids social situations or endures them with extreme anxiety or fear.
- The anxiety or fear exceeds the actual danger posed by the social situation.
- The fear or anxiety is persistent, as evidenced by lasting for at least 6 months.

Specifiers include:

- Performance only: if fear is limited to public performance

It is important to note that in children, anxiety should occur in peer settings and not just when interacting with adults; children may also express the fear by crying, tantrums, freezing, clinging, shrinking, or failing to speak in social situations.

## Panic Disorder

Panic disorder (F41.0) is diagnosed in people who have recurring, unexpected panic attacks, which are sudden surges of extreme fear or discomfort that reach a peak within minutes. During a panic attack, a person must experience:

- At least 4 of the following symptoms for diagnosis:
  - Palpitations, pounding heart, or increased heart rate
  - Sweating
  - Trembling or shaking
  - Feeling short of breath or like one is being smothered
  - The sensation of choking
  - Pain or discomfort in the chest
  - Nausea or upset stomach
  - Dizziness, unsteadiness, lightheadedness, or fainting
  - Chills or heat sensations
  - Paresthesia, which is the sensation of numbness or tingling
  - The experience of derealization, which is a feeling of unreality, or depersonalization, which is a sense of being detached from one's body
  - Fear of losing control or "going crazy"
  - Fear of death
- In addition to showing panic attack symptoms above, a person must have at least one month of ongoing concern or worry about having additional panic attacks or a noticeable, maladaptive change in behaviors, such as avoiding unfamiliar situations to reduce the risk of a panic attack.

## Panic Attack Specifier

A panic attack is not a diagnosis in and of itself, and it cannot be coded. However, panic attacks can occur with other anxiety disorders or with other mental health disorders, such as depressive disorders, posttraumatic stress disorder, and substance use disorders. They may also occur with some medical conditions. A panic attack specifier is used when a panic attack occurs with another disorder.

Panic attacks are sudden surges of extreme fear or discomfort that reach a peak within minutes.

During a panic attack, a person must experience at least 4 of the following symptoms for diagnosis:

- Sweating
- Palpitations, pounding heart, or increased heart rate
- Trembling or shaking
- Feeling short of breath or like one is being smothered
- The sensation of choking
- Pain or discomfort in the chest
- Nausea or upset stomach
- Dizziness, unsteadiness, lightheadedness, or fainting
- Chills or heat sensations
- Paresthesia, which is the sensation of numbness or tingling
- The experience of derealization, which is a feeling of unreality, or depersonalization, which is a sense of being detached from one's body
- Fear of losing control or "going crazy"
- Fear of death

## Agoraphobia

A person with agoraphobia (F40.00), experiences:

- Significant fear or anxiety about at least 2 of the following 5 situations:
- Use of public transportation, including automobiles, buses, trains, ships, and planes
- Being in open spaces, like parking lots or bridges

- Being in enclosed spaces, such as shops or cinemas
- Standing in line or being part of a crowd
- Being away from home alone

With agoraphobia, fear is related to thoughts that escape may be challenging, or help may be unavailable should one develop panic symptoms. While this is a separate diagnosis from panic disorder, a person can be diagnosed with both panic disorder and agoraphobia if their symptoms also meet the criteria for panic disorder.

## Generalized Anxiety Disorder

Generalized anxiety disorder (F41.1) is diagnosed when people have:

- Excessive fear and worry occur more days than not for 6 months or more and are related to numerous events or activities.
- Difficulty controlling their worry, and they must experience at least 3 of the following symptoms (only one in children), with at minimum some appearing more days than not for the previous 6 months:
  - A feeling of restlessness or being keyed up/on edge
  - Becoming fatigued easily
  - Having a difficult time concentrating or finding that one's mind goes blank
  - Irritability
  - Tension in muscles
  - Sleep problems, which can manifest as trouble falling or staying asleep or restless, unrefreshing sleep

## Other Anxiety Disorders

Other anxiety disorders in this chapter are as follows:

Substance/Medication-Induced Anxiety Disorder, which is diagnosed in people who have:

- Panic attacks or anxiety with evidence that symptoms developed during or soon after intoxication, withdrawal, or exposure to a substance or medication

Coding depends upon the specific substance and whether a person has a use disorder.

Specifiers include:

- With onset during intoxication
- With onset during withdrawal
- With onset after medication use.

Anxiety Disorder Due to Another Medical Condition (F06.4), which is diagnosed in people who have:

- Panic attacks or anxiety
- Evidence that the anxiety is a result of another medical condition.

Other Specified Anxiety Disorder (F41.8), which is diagnosed in people who have:

- Symptoms of an anxiety disorder but do not meet the full criteria for any of the anxiety disorders
- The clinician specifies the reason full criteria for another anxiety disorder are not met.

Unspecified Anxiety Disorder (F41.9), which is diagnosed in people who have:

- Symptoms of an anxiety disorder, but they do not meet full criteria for any of the anxiety disorders
- The clinician does not specify the reason full criteria for another anxiety disorder are not met.

CHAPTER 6

# Obsessive-Compulsive and Related Disorders

The obsessive-compulsive and related disorders chapter in the DSM-5-TR includes conditions like obsessive-compulsive disorder (OCD), body dysmorphic disorder, hoarding disorder, trichotillomania (hair-pulling disorder), and excoriation (skin-picking) disorder. The disorders in this chapter involve preoccupations and rituals that are excessive or persist beyond what is developmentally appropriate. Disorders in this chapter also involve repetitive behaviors, which can be body-focused.

The specific disorders in this chapter are detailed below.

## Obsessive-Compulsive Disorder

Obsessive-compulsive disorder (F42.2) is diagnosed in people who experience:

- Obsessions, compulsions, or both
- To meet diagnostic criteria, obsessions or compulsions must be time-consuming, as evidenced by taking more than one hour per day, or they must cause clinically significant distress or impairment in functioning.

People with OCD typically experience intrusive thoughts with these characteristics::

- Recurring, persistent thoughts, urges, or images that a person experiences as intrusive and unwanted and cause noticeable anxiety or distress in most individuals.

- Attempts to ignore or suppress the unwanted thoughts, urges, or images, or, alternatively, neutralize them with another thought or action, such as a compulsion.

Compulsions involve the following:
- Repetitive behaviors like handwashing, checking, or ordering, or mental acts like praying or counting that a person performs according to rigid rules in response to an obsession.
- Acts performed with the goal of preventing or reducing anxiety or distress or preventing a dreaded event or situation, even though they could not realistically neutralize or prevent these events or situations; alternatively, they are clearly excessive.

Specifiers include:
- With good or fair insight
- With poor insight
- With absent insight/delusional beliefs
- Tic-related

## Body Dysmorphic Disorder

Body dysmorphic disorder (F45.22) is diagnosed in people who are preoccupied with perceived defects or flaws in physical appearance, which are unobservable or only minor to others.

To be diagnosed, a person must meet the following criteria in addition to their persistent focus on perceived physical imperfections that others may not notice:
- Performing repetitive behaviors such as mirror checking, skin-picking, seeking reassurance, or excessive grooming at some point during the illness
- Experiencing clinically significant distress or dysfunction in daily life
- Evidence that the preoccupation is not related to concerns with body fat or weight in someone who meets the criteria for an eating disorder

Specifiers include:

- With muscle dysmorphia
- With good or fair insight
- With poor insight
- With absent insight/delusional beliefs

## Hoarding Disorder

People with hoarding disorder (F42.3) have persistent difficulty discarding or getting rid of possessions, even when possessions have little to no value. Diagnostic criteria for this disorder include:

- Experiencing a perceived need to save items and distress regarding discarding them
- Accumulating possessions to the extent that living areas are congested and cluttered which limits their use for intended purposes; third-party intervention may make living areas usable.

Specifiers include:

- With excessive acquisition
- With good or fair insight
- With poor insight
- With absent/delusional beliefs

## Trichotillomania (Hair-Pulling Disorder)

Trichotillomania (hair-pulling disorder) (F63.3) is diagnosed in people who:

- Repeatedly pull their hair, causing hair loss, and who also meet the following diagnostic criteria:
- Repeatedly attempting to reduce or stop hair-pulling
- Clinically significant distress or dysfunction resulting from hair-pulling
- Hair pulling or loss is not explained by another medical condition like a dermatological problem.

## Excoriation (Skin-Picking) Disorder

Excoriation (skin-picking) disorder (F42.4) is diagnosed in people who engage in:

- Repeated skin-picking, causing skin lesions, while also meeting the following criteria:
  - Repeatedly attempting to reduce or stop skin-picking
  - Clinically significant distress or dysfunction resulting from skin-picking
- Skin-picking is not explained by another medical condition like scabies
- Skin-picking is not explained by symptoms associated with another disorder like tactile hallucinations in psychotic disorder

## Other Obsessive-Compulsive and Related Disorders

Other obsessive-compulsive and related disorders in this chapter include:

Substance/Medication-Induced Obsessive-Compulsive and Related Disorder, which is diagnosed in people who show:

- Obsessions, compulsions, skin-picking, hair pulling, other body-focused repetitive behaviors, or other symptoms related to obsessive-compulsive and related disorders,
- There is evidence that symptoms developed during or soon after substance or medication intoxication, withdrawal, or exposure.

Coding depends upon the specific substance or medication and whether a person has a use disorder.

Obsessive-Compulsive and Related Disorder Due to Another Medical Condition (F06.8), which is diagnosed in people who show:

- Obsessions, compulsions, preoccupations with appearance, hoarding, skin-picking, hair pulling, other-body-focused repetitive behaviors, or other symptoms related to obsessive-compulsive and related disorders
- There is evidence that symptoms resulted directly from another medical condition.

Other Specified Obsessive-Compulsive and Related Disorder (F42.8), which is diagnosed in people who show:

- Symptoms related to obsessive-compulsive and related disorders, but they do not meet full diagnostic criteria for another disorder in this chapter
- The clinician does not specify the reason the person does not meet full diagnostic criteria for another disorder in this chapter.

# CHAPTER 7

# Trauma and Stressor-Related Disorders

The trauma and stressor-related disorders chapter of the DSM-5-TR is made up of disorders that arise after exposure to a traumatic or stressful event. These disorders are placed near surrounding chapters about related disorders, including anxiety disorders, obsessive-compulsive and related disorders, and dissociative disorders.

Sometimes, people develop fear or anxiety after a traumatic or stressful event. Other times, they show anhedonia and dysphoria, externalizing behaviors, or dissociation after a traumatic event. The trauma and stressor-related disorders chapter captures this presentation of symptoms following a traumatic or stressful event rather than disorders that present solely as fear or anxiety-based.

The specific disorders in this chapter are detailed below.

## Reactive Attachment Disorder

Reactive attachment disorder (F94.1) is diagnosed in children who have a developmental age of at least 9 months of age who show signs of a disturbance before the age of 5 years. It involves:

- Emotionally withdrawn and inhibited behavior toward adult caregivers, as evidenced by the child rarely or minimally seeking comfort when distressed and rarely or minimally responding to comfort from a caregiver when distressed.

Other criteria for diagnosis include:

- An ongoing social and emotional disturbance involving at least 2 of the following:
  - Being minimally socially and emotionally responsive to other people.
  - Showing little positive affect.
  - Episodes of unexplainable irritability, sadness, or fearfulness that appear even when interactions with adult caregivers are non-threatening.
- Evidence that the child has had extreme or insufficient care as a result of at least one of the following:
  - Social neglect or deprivation resulting from ongoing lack of emotional needs like comfort, stimulation, and affection being met by adult caregivers.
  - Recurring changes in primary caretakers that make it difficult to form stable attachments.
  - Being raised in unusual settings that makes it challenging to form selective attachments to caretakers.
- Evidence that extreme or insufficient care has caused the behavior disturbance and criteria are not present for autism spectrum disorder.

Specifiers include:

- Persistent (given when the disorder is present more than 12 months)
- Severe (given when a child shows all symptoms at relatively high levels)

## Disinhibited Social Engagement Disorder

Disinhibited Social Engagement Disorder (F94.2) is diagnosed in children who have a developmental age of at least 9 month and demonstrate a pattern of actively approaching and interacting with unfamiliar adults. To be diagnosed, a child must show:

- At least 2 of the following symptoms:
  - Reduced or lack of shyness when approaching and interacting with unfamiliar adults.

- Being overly familiar in verbal or physical behavior in a way that is not consistent with cultural norms and age-appropriate social boundaries.
- Venturing away from an adult caregiver with little to no checking back with the caretaker, even in an unfamiliar setting.
- Being willing to go off with an unfamiliar adult with little to no hesitation.

Additional diagnostic criteria include:

- The symptoms above are not limited to impulsivity and include evidence of social disinhibition.
- The child has a history of extreme or insufficient care, which is evidenced by one or more of the following:
  - Social neglect or deprivation resulting from ongoing lack of emotional needs like comfort, stimulation, and affection being met by adult caregivers.
  - Recurring changes in primary caretakers that make it difficult to form stable attachments.
  - Being raised in unusual settings that make it difficult to form selective attachments to caretakers.
- Extreme or insufficient care is responsible for the behavioral disturbance.

Specifiers include:

- Persistent (given when the disorder is present more than 12 months)
- Severe (when a child shows all symptoms at relatively high levels)

## Posttraumatic Stress Disorder

Posttraumatic stress disorder (F43.10) is diagnosed in individuals who are exposed to a traumatic event. Diagnostic criteria are different for children younger than 6 years old.

The diagnostic criteria for adults, adolescents, and children older than 6 years of age are as follows:

- Experiencing a traumatic event which involves actual or threatened death, serious injury, or sexual violence in at least one of the following ways, which does not apply to exposure through television or electronic media unless it is work-related:
  - Directly experiencing a traumatic event.
  - Witnessing a traumatic event occur to others in person.
  - Learning that a traumatic event occurred to a close family member or friend; if the event is acute or threatened death, it must be violent or accidental.
  - Working in a role such as that of a first responder or police officer and experiencing recurring extreme exposure to negative details of traumatic events.
- Showing at least one of the following intrusion symptoms related to the traumatic event:
  - Ongoing memories that are viewed as intrusive and unwanted and lead to stress; in children this can involve playing in a way in which features of the event are demonstrated.
  - Ongoing dreams with emotions or content related to the trauma, which may occur as frightening dreams with ambiguous content in children.
  - Dissociative reactions like flashbacks during which the person acts or feels like the traumatic event is occurring again; this can range in severity and involve a total loss of awareness of surroundings, and in children, it can be reenacting trauma in the context of play.
  - Showing severe or lasting psychological distress when exposed to stimuli, either internal or external, that incite memories of the traumatic event.
  - Experiencing noticeable physiological reactions to cues, whether internal or external, that trigger trauma.
- Persistently avoiding stimuli linked to the traumatic event, which manifests as one or both of the following:
  - Avoiding or attempting to avoid distressing memories, thoughts, or feelings related to the trauma
  - Avoiding or attempting to avoid people, places, conversations, and other outward reminders related to the trauma.

- Negative changes involving cognition in mood, which can be linked to the trauma, as evidenced by at least two of the following:
  - Failing to remember important features of the traumatic event as a result of factors like dissociative amnesia.
  - Ongoing, extreme negative beliefs or expectations related to oneself, other people, or the world; could involve the belief that no one can be trusted, for instance.
  - Ongoing distorted cognitions centering on the cause or outcomes of the traumatic event, such as the belief that one is to blame for the trauma.
- Being persistently negative in emotionality, such as showing persistent anger, horror, guilt, or shame.
  - Significantly reducing participation in usual activities or showing lack of interest in such activities.
  - Feeling cut off from other people.
  - Having an ongoing inability to experience positive emotionality, such as happiness or joy.
- Noticeable changes in arousal levels and reactivity linked to the traumatic event, manifesting as at least two of the following symptoms:
  - Irritability and outbursts of anger, which typically involve physical or verbal aggression toward people or objects and occur with little to no provocation.
  - Engaging in behavior that is reckless or harmful to oneself.
  - Being constantly on the lookout for threats.
  - Startling easily.
  - Difficulty concentrating.
  - Trouble falling or staying asleep, or restless sleep
- A disturbance lasting more than one month.

Specifiers include:

- With dissociative symptoms (used when a person experiences persistent depersonalization or derealization)
- With delayed expression (used when full diagnostic criteria are not met until at least 6 months following the traumatic event)

The diagnostic criteria for children aged 6 years and younger are as follows:
- Experiencing a traumatic event which involves actual or threatened death, serious injury, or sexual violence in at least one of the following ways:
  - Directly experiencing a traumatic event.
  - Observing an event occur to others in person, especially when the event happens to primary caretakers.
  - Learning about a traumatic event occurring to a parent or caretaker.
- Experiencing at least one of the following intrusion symptoms related to the traumatic event:
  - Repeated, involuntary, and intrusive memories of the traumatic event that cause distress; this may present as memories expressed through play, and they may not appear distressing.
  - Repeated distressing dreams involving content or affect related to the traumatic event; it may be impossible to determine that the frightening content is associated with the traumatic event.
  - Dissociative reactions like flashbacks during which the person acts or feels like the traumatic event is occurring again; this can range in severity and involve a total loss of awareness of surroundings, and it can appear as reenactment during play.
  - Showing intense or prolonged psychological distress when exposed to stimuli, either internal or external, that remind one of the traumatic event or an aspect of it.
  - Experiencing noticeable physiological reactions to internal or external cues related to the trauma.
  - Experiencing either one symptom related to persistent avoidance or one symptom related to negative alterations in cognition and mood linked to the traumatic event.
- Persistent avoidance can manifest as:
  - Avoiding or attempting to avoid distressing memories, thoughts, or feelings related to the trauma.
  - Avoiding or attempting to avoid people, places, conversations, and other external reminders related to the trauma.
- Negative alterations in cognitions can manifest as:
  - Increase in negative emotional states like fear, guilt, or sadness

- Lack of interest or participation in usual activities, including play
- Socially withdrawing
- Reduced expression of positive emotions, which persists
- Alterations in arousal and reactivity related to the traumatic event, manifesting as at least 2 of the following symptoms:
  - Irritability and outbursts of anger with little to no provocation, which are often expressed as verbal or physical aggression toward people or objects, can include extreme temper tantrums
  - Hypervigilance
  - Extreme startle response
  - Difficulty concentrating
  - Trouble falling or staying asleep, or restless sleep
- A disturbance lasting more than one month.

Specifiers include:

- With dissociative symptoms (used when a person experiences persistent depersonalization or derealization)
- With delayed expression (used when full diagnostic criteria are not met until at least 6 months following the traumatic event)

## Acute Stress Disorder

Acute stress disorder (F43.0) is diagnosed in people who are:

- Exposed to actual or threatening death, serious injury, or sexual violence through one of the following means:
  - Directly experiencing a traumatic event
  - Witnessing a traumatic event occur to others in person
  - Learning of a traumatic event occurring to a close family member or friend; if the event is acute or threatened death, it must be violent or accidental.
  - Working in a role such as that of a first responder or police officer and experiencing recurring extreme exposure to negative details of traumatic events.

In addition to the above, a person must show at least 9 of the following symptoms from any of the 5 categories below:

- **Intrusion Symptoms**
    - Repeated, involuntary, and intrusive memories of the traumatic event that cause distress; in children, this can involve repetitive play in which themes or aspects of the traumatic event are expressed.
    - Repeated distressing dreams involving content or affect related to the traumatic event; this may manifest as frightening dreams with unrecognizable content in children.
    - Dissociative reactions like flashbacks during which the person acts or feels like the traumatic event is occurring again; this can range in severity and involve a total loss of awareness of surroundings, and in children, it can include trauma reenactment during play.
    - Showing either intense or prolonged psychological distress or noticeable physiological reactions in reaction to internal or external cues related to the trauma.
- **Negative Mood**
    - Persistently showing an inability to experience positive emotions like happiness, satisfaction, or feelings of love
- **Dissociative Symptoms**
    - Showing an altered sense of reality related to surroundings or oneself
    - Being unable to remember an important aspect of the trauma usually experienced as dissociative amnesia
- **Avoidance Symptoms**
    - Making attempts to avoid distressing memories, thoughts, or feelings related to the traumatic event
    - Making attempts to avoid external reminders, such as people, places, or conversations that bring up distressing memories, thoughts, or feelings related to the trauma
- **Arousal Symptoms**
    - Difficulty falling or staying asleep or restless sleep
    - Irritability and angry outbursts with little to no provocation, usually expressed as physical or verbal aggression to people or objects

- Hypervigilant behavior
- Difficulty with concentration
- Extreme startle response
* Other diagnostic criteria include:
  - Disturbance lasting at least 3 days but not more than one month after trauma exposure

## Adjustment Disorders

Adjustment disorders are diagnosed in people who have emotional or behavioral symptoms in response to an identifiable stressor that occurred within the last 3 months.

To be diagnosed, a person must meet at least one of the following criteria:

- Significant distress that is out of proportion given the severity or intensity of the stressor, in light of the external environment and cultural factors influencing the severity and presentation of symptoms.
- Significant dysfunction in social, occupational, or other important settings.

Additional criteria include:

- The disturbance cannot be better explained by another mental disorder
- It is not an exacerbation of an existing mental disorder.
- The symptoms also must not represent typical bereavement or prolonged grief disorder
- Symptoms persist for no longer than 6 additional months after the stressor and its consequences have terminated.

Specifiers and coding include:

- With depressed mood: F43.21
- With anxiety: F43.22
- With mixed anxiety and depressed mood: F43.23
- With disturbance of conduct: F43.24
- With mixed disturbance of emotions and conduct: F43.25
- Unspecified: F43.20

Other specifiers include:

- Acute
- Persistent (chronic)

## Prolonged Grief Disorder

Prolonged grief disorder (F43.8) is diagnosed in people who have experienced the death of a person close to them 12 or more months ago (or 6 or more months ago in children and adolescents). In addition to experiencing the death, the person will also show:

- Intense yearning or longing for the deceased person or preoccupation with thoughts or memories of the deceased person.
- This persistent grief response is present most days for at least the previous month.
- Additional criteria for diagnosis include showing at least 3 of the following symptoms:
    - Disruption in identity, which may manifest as feeling part of oneself has died
    - Noticeable sense of disbelief about the death
    - Avoiding reminders that the person has died
    - Intense emotional pain, which may include anger, bitterness, or sorrow
    - Trouble reintegrating into usual relationships and activities after the death
    - Emotional numbness
    - Feeling life is meaningless
    - Extreme loneliness

## Other Trauma and Stressor-Related Disorders

Other trauma and stressor-related disorders in this chapter include:

Other Specified Trauma and Stressor-Related Disorder (F43.8), diagnosed in people who have:

- Symptoms characteristic of a trauma and stressor- related disorder,

- Symptoms do not meet the full criteria for another disorder in this chapter
- The clinician specifies the reason why symptoms do not meet the criteria for another disorder in this chapter.

Unspecified Trauma and Stressor-Related Disorder (F43.9), diagnosed in people who have:

- Symptoms characteristic of a trauma and stressor-related disorder
- Symptoms do not meet full criteria for another disorder in this chapter
- The clinician does not specify the reason why symptoms do not meet the full criteria for another disorder in this chapter.

# CHAPTER 8

# Dissociative Disorders

The dissociative disorders chapter in the DSM-5-TR includes conditions that involve disruption or discontinuity in the typical integration of consciousness, memory, identity, emotion, perception, body representation, motor control, and behavior. These disorders usually occur in response to psychologically traumatizing experiences. These disorders are placed next to the trauma and stressor-related disorders chapter, given their close association.

Dissociative symptoms involve intrusions into awareness and behavior, which causes loss of continuity in subjective experiences. Dissociative symptoms can be positive, involving identity division, depersonalization, derealization, and inability to control mental functions. They can also be negative, involving symptoms like amnesia.

The specific disorders in this chapter are detailed below.

## Dissociative Identity Disorder

Dissociative identity disorder (F44.81) is diagnosed in people who have:

- At least 2 distinct personality states, which may be described as possession in some cultures.
- The distinct personality states in this disorder result in a noticeable discontinuity in the sense of self and agency, in addition to alterations in mood, behavior, consciousness, memory, perception, cognition, or sensory-motor functioning.

Other diagnostic criteria for dissociative identity disorder include:

- Recurring gaps in memory of daily events, important personal information, or traumatic events that are not simply ordinary forgetting
- Clinically significant distress or dysfunction in social, occupational, or other important areas
- Disturbance is not explained by normal or broadly accepted cultural or religious customs, and in children, it is not the result of imaginary playmates or fantasy play

## Dissociative Amnesia

Dissociative amnesia (F44.0) is diagnosed in people who are unable to recall autobiographical information, usually that which is traumatic or stressful. With dissociative amnesia, symptoms are not simply ordinary forgetting. This type of amnesia typically involves localized or selective amnesia related to a specific event or events or generalized amnesia related to identity and life history.

Other diagnostic criteria for dissociative amnesia include:

- Clinically significant distress or dysfunction in social, occupational, or other important areas.
- The disturbance is not explained by the effects of a substance or by a neurological or other medical condition.
- The disturbance is not better explained by dissociative identity disorder, posttraumatic stress disorder, somatic symptom disorder, or major or mild neurocognitive disorder.

Specifiers include:

- With dissociative fugue, which results in coding change to F44.1. This is diagnosed in people who show evidence of apparently purposeful travel or bewildered wandering linked to identity amnesia.

## Depersonalization/Derealization Disorder

People with depersonalization/derealization disorder (F48.1) show an ongoing pattern of depersonalization, derealization, or both. Depersonalization involves the experience of being unreal or detached from one's thoughts, feelings, sensations, body, or actions. Derealization involves the experience of being unreal or detached from one's surroundings.

Additional diagnostic criteria for depersonalization/derealization disorder include:

- Intact reality testing during experiences of depersonalization or derealization.
- Clinically significant distress or dysfunction in social, occupational, or other important areas.
- The disturbance is not attributed to the effects of a substance and is not better explained by another mental health disorder.

## Other Specified Dissociative Disorder

Other specified dissociative disorder (F44.89) is diagnosed in people who have:

- Symptoms characteristic of a dissociative disorder that do not meet full criteria for another disorder in this chapter
- The clinician specifies the reason full criteria for another dissociative disorder are not met.

## Unspecified Dissociative Disorder

Unspecified dissociative disorder (F44.9), is diagnosed in people who have:

- Symptoms characteristic of a dissociative disorder but that do not meet full criteria for another disorder in this chapter
- The clinician does not specify the reason full criteria for another dissociative disorder are not met.

# CHAPTER 9

# Somatic Symptom and Related Disorders

The somatic symptom and related disorders chapter in the DSM-5-TR includes mental health disorders associated with somatic symptoms or illness anxiety that result in significant distress and impairment. These disorders are more likely to be found in people presenting for treatment in primary care or medical settings rather than psychiatric settings.

The primary disorder in this chapter is somatic symptom disorder, which involves distressing somatic symptoms paired with abnormal thoughts, feelings, and behaviors related to these symptoms.

The specific disorders in this chapter are detailed below.

## Somatic Symptom Disorder

Somatic symptom disorder (F45.1) is diagnosed in people who have:

- At least one somatic symptom that is distressing and causes significant disruption in daily life.
- The somatic symptom(s) may not continuously be present, but for diagnosis, the somatic state persists, usually for more than 6 months.
- To meet the criteria for diagnosis, a person must present with excessive thoughts, feelings, or behaviors related to somatic symptoms or related health concerns, as evidenced by one or more of the following symptoms:
  - Persistent, disproportionate thoughts related to the seriousness of the symptoms.

- Persistently high anxiety levels are related to the symptoms or one's health.
- Devoting excessive time or energy to symptoms or health concerns.

Specifiers include:

- With predominant pain
- Persistent
- Mild (only one symptom of excessive thoughts, feelings, or behaviors met)
- Moderate (2 or more symptoms of excessive thoughts, feelings, or behaviors met)
- Severe (2 or more symptoms of excessive thoughts, feelings, or behaviors met, combined with multiple somatic complaints or one very severe somatic complaint)

## Illness Anxiety Disorder

Illness anxiety disorder (F45.21) is diagnosed in people who are preoccupied with having or acquiring a serious illness. People with illness anxiety disorder do not have somatic symptoms, or they have only mild somatic complaints. If they have a medical condition or are at high risk for developing one, their preoccupation is clearly excessive or disproportionate to the risk or danger.

Other criteria for diagnosis include:

- High anxiety levels related to health, paired with becoming easily alarmed about one's health status.
- Excessive performance of health-related behaviors, such as checking for signs of illness, or, alternatively, maladaptive avoidance, which may involve avoiding doctor's appointments).
- Showing at least 6 months of illness preoccupation, but the specific illness causing fear may change during that time.
- Preoccupation related to illness is not better explained by another mental disorder.

Specifiers include:

- Care-seeking type
- Care-avoidant type

## Functional Neurological Symptom Disorder (Conversion Disorder)

Functional neurological symptom disorder, also called conversion disorder, is diagnosed in people who have:

- At least one symptom of altered motor or sensory function.
- In people with this disorder, clinical findings show an incompatibility between the symptoms and recognized neurological or medical conditions.
- Finally, symptoms must not be bettered explained by another mental or medical disorder, and they must cause clinically significant distress or dysfunction.

Coding is based on symptom type, as follows:

- F44.4: With weakness or paralysis
- F44.4: With abnormal movements
- F44.4: With swallowing symptoms
- F44.4: With speech symptom
- F44.5: With attacks or seizures
- F44.6: With anesthesia or sensory loss
- F44.6: With special sensory symptom
- F44.7: With mixed symptoms

Other specifiers include:

- Acute episode: symptoms present for less than 6 months
- Persistent: symptoms are present for 6 months or more.
- With psychological stressor
- Without psychological stressor

## Psychological Factors Affecting Other Medical Conditions

This symptom or condition (F54) is used when a person has a medical symptom or condition aside from a mental disorder, and psychological or behavioral factors negatively affect the medical condition, as evidenced by one of the following:

- Affect the medical condition's course, as indicated by a close temporal association between the factors and development, worsening of, or delayed recovery from the medical condition
- Impede treatment of the medical condition
- Are a well-established health risk for the person
- Affect underlying pathophysiology, which precipitates or worsens symptoms or necessitates medical intervention

Specifiers include:

- Mild (increases medical risk)
- Moderate (aggravates underlying medical condition)
- Severe (results in hospitalization or emergency room visit)
- Extreme (results in severe, life-threatening risks)

## Factitious Disorder

There are 2 types of factitious disorder: factitious disorder imposed on self (F68.10) and factitious disorder imposed on other (F68.A).

The diagnostic criteria for factitious disorder imposed on self include:

- Falsifying physical or psychological signs and symptoms or deceptively inducing injury or disease.
- Presenting oneself to other people as ill, impaired, or injured.
- Deceptive behavior continues even without obvious outside rewards.
- Another mental disorder, like a delusional disorder or a psychotic disorder, does not explain the symptoms.

Specifiers include:

- Single episodes
- Recurrent episodes

The diagnostic criteria for factitious disorder imposed on another include:

- Falsifying physical or psychological signs and symptoms or inducing injury or disease in another person in a deceptive way.
- Presenting another person, who is the victim, to others as ill, impaired, or injured.
- Deceptive behavior continues even without obvious outside rewards.
- Another mental disorder, like a delusional disorder or a psychotic disorder, does not explain the symptoms.

The perpetrator of the above, rather than the victim, is given the diagnosis.

Specifiers include:

- Single episode
- Recurrent episodes

## Other Specified Somatic Symptom and Related Disorder

Other specified somatic symptom and related disorder (F45.8) is diagnosed in people who have:

- Symptoms characteristic of a somatic symptom and related disorder
- Symptoms do not meet the full criteria for another disorder in this chapter
- The clinician specifies why the person does not meet full diagnostic criteria for another disorder.

## Unspecified Somatic Symptom and Related Disorder

Unspecified somatic symptom and related disorder (F45.9) is diagnosed in people who have:

- Symptoms characteristic of a somatic symptom and related disorder
- Symptoms do not meet full criteria for another disorder in this chapter
- The clinician does not specify the reason the person does not meet full diagnostic criteria for another disorder.

# CHAPTER 10

# Feeding and Eating Disorders

The feeding and eating disorders chapter in the DSM-5-TR includes disorders that involve a persistent disturbance in eating or eating-related behaviors that cause altered consumption or absorption of food and that significantly harm health or psychosocial functioning. It is worth noting that the diagnostic criteria for anorexia nervosa, bulimia nervosa, and binge-eating disorder are mutually exclusive, so a person cannot be diagnosed with more than one of these disorders.

The specific disorders in this chapter are detailed below.

## PICA

Pica is diagnosed in people who persistently eat nonnutritive, nonfood substances for a period lasting at least 1 month.

Additional diagnostic criteria for pica include:

- The eating behavior is not appropriate to the person's developmental level.
- The eating behavior isn't related to a cultural or social practice.
- The eating behavior warrants additional clinical attention if it occurs in the context of another condition, like intellectual developmental disorder, autism spectrum disorder, or schizophrenia.

Coding is as follows:

- F98.3: Children
- F50.89: Adults

Chapter 10: Feeding and Eating Disorders

Specifiers include:

- In remission

## Rumination Disorder

Rumination disorder (F98.21) is diagnosed in people who repeatedly regurgitate food over a period lasting 1 month or more. The food may be re-chewed, re-swallowed, or spit out after regurgitation.

Other criteria for diagnosis include:

- Repeated regurgitation cannot be explained by a gastrointestinal or other medical condition.
- The disturbed eating does not occur in the context of one of these eating disorders: anorexia nervosa, bulimia nervosa, binge-eating disorder, or avoidant/restrictive food intake disorder.
- The eating behavior warrants additional clinical attention if it occurs in the context of another condition, like intellectual developmental disorder, autism spectrum disorder, or schizophrenia.

Specifiers include:

- In remission

## Avoidant/Restrictive Food Intake Disorder

Avoidant/restrictive food intake disorder (F50.82) is diagnosed in people who show a disturbance in feeding or eating, characterized by:

- A lack of interest in food, avoidance of food based on its sensory characteristics, or concern that food will bring negative consequences.
- For diagnosis, a person must show at least one of the following symptoms:
- Losing a significant amount of weight or missing expected weight gain or growth targets during childhood.
- Showing deficiencies in key nutrients.
- Requiring enteral feeding or nutrition supplements due to lack of food intake.
- Showing significant psychosocial dysfunction.

Other criteria include:

- Eating disturbance cannot be explained by a lack of food availability or a cultural practice.
- Eating disturbance cannot occur only in the context of anorexia nervosa or bulimia nervosa, and there cannot be evidence of a disturbed perception of one's body weight or shape.
- The eating disturbance is not explained by a medical or mental health disorder.

Specifiers include:

- In remission

## Anorexia Nervosa

Anorexia nervosa is diagnosed in people who restrict food intake below typical energy requirements, causing significantly low body weight, which is less than what is minimally normal or expected.

Other diagnostic criteria include:

- Extreme fear of gaining weight or becoming fat, or ongoing behavior that prevents weight gain despite being at a significantly low weight.
- Experiencing one's body weight or shape in a disturbed fashion, placing extreme importance on body weight or shape when evaluating oneself, or denying the seriousness of the low body weight.

Coding occurs according to the following specifiers:

- F50.01: Restricting type
- F50.02: Binge-eating/purging subtype (includes purging subtype even when binge-eating is absent)

Additional specifiers include:

- In partial remission
- In full remission
- Mild: BMI greater than or equal to 17

- Moderate: BMI of 16-16.99
- Severe: BMI of 15-15.99
- Extreme: BMI under 15

## Bulimia Nervosa

Bulimia nervosa (F50.2) is diagnosed in people who recurrently engage in binge eating.

The following criteria must be met to make the diagnosis:

- Within a discrete period, eating more food than most people would consume in a similar period and circumstances.
- Losing control of eating during the binge episode.

In addition to meeting the criteria for a binge, a person must meet the following diagnostic criteria:

- Engaging in repeated compensatory behaviors, such as self-induced vomiting, laxative or diuretic misuse, fasting, or excessive exercise to avoid weight gain.
- Binging and engaging in compensatory behaviors at least once weekly for 3 months.
- Placing extreme importance on body weight or shape when evaluating oneself.
- Evidence that the disturbance does not occur only in the context of anorexia.

Specifiers include:

- In partial remission
- In full remission
- Mild: average of 1-3 episodes per week
- Moderate: average of 4-7 episodes per week
- Severe: average of 8-13 episodes per week
- Extreme: average of 14 or more episodes per week

## Binge-Eating Disorder

Binge-eating disorder (F50.81) is diagnosed in people who engage in repeated episodes of binge eating.

Diagnosis requires both of the following criteria to be met:

- Within a discrete period, eating a larger amount of food than most people would consume in a similar time period and circumstances.
- Losing control of eating during the binge episode.
- During the binge eating episodes, showing 3 or more of the following symptoms:
- Consuming food more quickly than what is considered typical.
    - Consuming food until reaching a state of discomfort from fullness.
    - Consuming large amounts of food despite lack of hunger.
    - Eating alone because of embarrassment about consuming so much food.
    - Following the binge, feeling guilty, depressed, or disgusted.

Additional diagnostic criteria include:

- Noticeable distress related to binge eating.
- Binging an average of at least one time weekly for 3 months.
- Lack of compensatory behavior after binges, which would be seen in bulimia nervosa, and the person does not have anorexia nervosa.

Specifiers include:

- In partial remission.
- In full remission.
- Mild: 1-3 binge episodes weekly
- Moderate: 4-7 binge episodes weekly
- Severe: 8-13 binge episodes weekly
- Extreme: 14 or more binge episodes weekly

## Other Specified Feeding or Eating Disorder

Other specified feeding or eating disorder (F50.89) is diagnosed in people who have:

- Symptoms characteristic of a feeding or eating disorder
- Symptoms do not meet the full criteria for any of the other disorders in this chapter
- The clinician specifies the reason the person does not meet full criteria for another disorder.

## Unspecified Feeding or Eating Disorder

Unspecified feeding or eating disorder (F50.9) is diagnosed in people who have:

- Symptoms characteristic of a feeding or eating disorder
- Symptoms do not meet full criteria for any of the other disorders in this chapter
- The clinician does not specify the reason the person does not meet full diagnostic criteria for another disorder.

# CHAPTER 11

# Elimination Disorders

The elimination disorders chapter in the DSM-5-TR is a short chapter that includes disorders that involve inappropriate elimination of urine and feces. These disorders are typically diagnosed in children and adolescents.

The specific disorders in this chapter are detailed below.

## Enuresis

Enuresis (F98.0) is diagnosed in individuals who either involuntarily or intentionally void urine into the bed or clothing.

Other criteria for diagnosis include:

- Behavior is clinically significant, evidenced by occurring at least twice a week for 3 consecutive months or substantial impairment in functioning.
- The person being diagnosed is at least 5 years of age.
- The behavior is not explained by the physiological effects of a substance or by another medical condition.

Specifiers include:

- Nocturnal only
- Diurnal only
- Nocturnal and diurnal

## Encopresis

Encopresis (F98.1) is diagnosed in individuals who either involuntarily or intentionally pass feces into inappropriate places.

Other criteria for diagnosis include:

- Incidents occur at least once monthly for at least 3 months.
- The person being diagnosed is at least 4 years of age.
- The behavior is not explained by the physiological effects of a substance or by another medical condition, except for one causing constipation.

Specifiers include:

- With constipation and overflow incontinence
- Without constipation and overflow incontinence

## Other Specified Elimination Disorder

Other specified elimination disorder is diagnosed in people who have:

- Symptoms characteristic of an elimination disorder
- Symptoms do not meet the full criteria for any of the other disorders in this chapter
- The clinician specifies the reason the person does not meet the full criteria for another disorder.

Coded as:

- N39.498 if it involves urinary symptoms
- R15.9 if it involves fecal symptoms.

## Unspecified Elimination Disorder

Unspecified elimination disorder is diagnosed in people who have:

- Symptoms characteristic of an elimination disorder
- Symptoms do not meet full criteria for any of the other disorders in this chapter
- The clinician does not specify the reason the person does not meet full diagnostic criteria for another disorder.

Coded as:

- R32 if it involves urinary symptoms
- R15.9 if it involves fecal symptoms.

# CHAPTER 12

# Sleep-Wake Disorders

The sleep-wake disorders chapter in the DSM-5-TR includes disorders that both mental health and general medical clinicians may diagnose. These disorders all create daytime distress and impairment due to sleep complaints. It is noted that individuals diagnosed with a disorder in this chapter may require a referral to a sleep medicine specialist.

The specific disorders in this chapter are detailed below.

## Insomnia Disorder

Individuals with insomnia disorder (F51.01) present with a primary complaint of:

- Dissatisfying sleep quality or quantity, which causes at least one of the following symptoms:
    - Difficulty falling asleep, which in children can present as difficulty falling asleep without a caretaking intervention
    - Difficulty staying asleep can present as frequent awakenings or difficulty falling back to sleep after awakenings; in children, this can manifest as difficulty falling back asleep without caretaker intervention.
    - Waking early in the morning and being unable to fall asleep.

Other essential criteria for diagnosis include:

- Clinically significant disturbance or impairment resulting from the sleep problem.
- Sleep difficulty occurs at least 3 nights weekly.
- Sleep difficulty lasting at least 3 months.

Specifiers include:

- With mental disorder
- With medical condition
- With another sleep disorder
- Episodic
- Persistent
- Recurrent

## Hypersomnolence Disorder

Hypersomnolence disorder (F51.11) is diagnosed in people who have:

- Excessive sleepiness, per self-report, despite sleeping at least 7 hours nightly.
- At least one of the following symptoms:
  - Recurrent sleep periods or lapses into sleep within the course of a day.
  - Non-restorative main sleep episode lasting more than 9 hours per day.
  - Difficulty becoming fully awake after abrupt awakening.

Other important criteria for diagnosis include:

- Hypersomnolence occurs at least 3 times weekly for 3 months.
- Clinically significant disturbance or impairment arising from hypersomnolence.

Specifiers include:

- With mental disorder
- With medical condition
- With another sleep disorder
- Acute
- Subacute
- Persistent

## Narcolepsy

Narcolepsy is diagnosed in people who experience:

- Repeated periods of irresistible need to sleep, lapsing into sleep, or need to nap within the same day.
- These episodes must occur at least 3 times weekly over the last 3 months.
- At least one of the following must be present for diagnosis:
    - Episodes of cataplexy occurring at least a few times monthly in people with longstanding disease, which can include brief episodes of sudden bilateral loss of muscle tone that are precipitated by laughing or joking and during which consciousness is maintained.
    - In children and those within 6 months of the onset of the disorder, spontaneous grimaces or jaw-opening episodes involve tongue thrusting or global hypotonia with no emotional triggers.
    - Hypocretin deficiency is measured using cerebrospinal fluid hypocretin-1 immunoreactivity values, which are not observed in the context of acute brain injury, inflammation, or infection.
    - Nocturnal sleep polysomnography shows rapid eye movement (REM) sleep latency under or equal to 15 minutes or multiple sleep latency tests indicating a mean sleep latency of under or equal to 8 minutes and 2 or more sleep-onset REM periods.

Coding is as follows, according to specifiers:

- G47.411: Narcolepsy with cataplexy or hypocretin deficiency (type 1)
- G47.419: Narcolepsy without cataplexy and either without hypocretin deficiency or hypocretin unmeasured (type 2)
- G47.421: Narcolepsy with cataplexy or hypocretin deficiency due to a medical condition
- G47.429: Narcolepsy without cataplexy and without hypocretin deficiency due to a medical condition

Additional specifiers include:

- Mild
- Moderate
- Severe

## Obstructive Sleep Hypopnea

Obstructive sleep hypopnea (G47.33) is diagnosed in people who have evidence of one of the following:

- Evidence per polysomnography of at least 5 obstructive sleep apneas or hypopneas per hour during sleep and one of the following symptoms:
    - Nighttime breathing disturbance, which involves snoring, snorting/gasping, or breathing pauses during sleep, which is not due to another mental disorder or medical condition.
    - Daytime sleepiness, fatigue, and unrefreshing sleep, despite having sufficient sleep opportunity, which is not due to another mental disorder or medical condition.

    OR

- Evidence per polysomnography of at least 15 obstructive apneas or hypopneas per hour during sleep, regardless of co-occurring symptoms.

Specifiers include:

- Mild
- Moderate
- Severe

## Central Sleep Apnea

Central sleep apnea is diagnosed in people who have:

- Evidence via polysomnography of at least 5 central sleep apneas per hour when sleeping
- The breathing problem is not better explained by another sleep disorder.

Coding occurs according to the following specifiers:

- G47.31: Idiopathic central sleep apnea
- R06.3: Cheyne-Stokes breathing
- G47.37: Central sleep apnea comorbid with opioid use.
- This disorder is also specified according to severity.

## Sleep-Related Hypoventilation

Sleep-related hypoventilation is diagnosed in people who show:

- Periods of decreased respiration linked to elevated carbon dioxide levels
- The disturbance is not attributed to another sleep disorder.

Coding occurs according to the following specifiers:

- G47.34: Idiopathic hypoventilation
- G47.35: Congenital central alveolar hypoventilation
- G47.36: Comorbid sleep-related hypoventilation
- This disorder is also specified according to severity.

## Circadian Rhythm Sleep-Wake Disorders

Circadian rhythm sleep-wake disorders are diagnosed in people who experience:

- An ongoing or repeated pattern of disrupted sleep due to changes in the circadian system or misalignment between the natural circadian rhythm and the sleep-wake schedule a person follows.
- To be diagnosed, a person must experience excessive daytime sleepiness, insomnia, or both
- They must experience clinically significant distress or impairment in daily functioning.

Coding occurs according to the following specifiers:

- G47.21: Delayed sleep phase type
  - Further specified as familial or overlapping with non-24-hour sleep-wake type
- G47.22: Advanced sleep phase type
  - Can be further specified as familial
- G47.23: Irregular sleep-wake type
- G47.24: Non-24-hour sleep-wake type
- G47.26: Shift work type
- G47.20: Unspecific type

The disorder can be further specified as follows:

- Episodic
- Persistent
- Recurrent

## Parasomnias

Parasomnias are diagnosed in people who experience abnormal behavior or experiential or physiological events during sleep, with specific sleep stages or during sleep-wake transitions. These disorders are described below in more detail.

## Non-rapid Eye Movement Sleep Arousal Disorders

These disorders are diagnosed in people who:

- Repeatedly experience an incomplete awakening from sleep, typically during the first third of the sleep episode, along with either of the following:
  - Sleepwalking
  - Sleep terrors

Additional important criteria for diagnosis include:

- Little to no dream imagery is remembered.
- There is amnesia for the episodes.

Specifiers and coding are as follows:

- F51.3: Sleepwalking type
  - Further specified as with sleep-related eating or with sleep-related sexual behavior (sexsomnia)
- F51.4: Sleep terror type

## Nightmare Disorder

Nightmare disorder (F51.5) is diagnosed in people who have:

- Repeated experiences of extended, extremely unpleasant, and well-recalled dreams that typically involve efforts to avoid threats to survival, security, or physical integrity. These episodes usually occur during the second half of sleep.

Other important criteria for diagnosis include:

- When awakening from unpleasant dreams, the person quickly becomes alert and oriented.
- Disturbance results in clinically significant distress or impairment in functioning.
- Nightmare symptoms are not explained by the physiological effects of a substance and are not attributable to coexisting mental disorders or medical conditions.

Specifiers include:

- During sleep onset
- With mental disorder, which include substance use disorders
- With medical condition
- With another sleep disorder
- Acute: Duration is less than one month
- Subacute: Greater than one month but less than 6 months.
- Persistent: 6 months or more
- Mild: Less than one episode per week
- Moderate: Not nightly but one or more episodes weekly
- Severe: Nightly episodes

## Rapid Eye Movement Sleep Behavior Disorder

Rapid eye movement sleep behavior disorder (G47.52) is diagnosed in people who experience:

- Repeated arousals during sleep linked to vocalization or complex motor behaviors.

Additional criteria for diagnosis include:
- Behaviors arise during rapid eye movement (REM sleep), meaning they typically occur more than 90 minutes after sleep onset, are more frequent later in the sleep period, and are uncommon during daytime naps.
- The person is entirely awake and oriented, rather than confused or disoriented, upon awakening from these episodes.
- The person experiences either REM sleep without atonia per polysomnographic recording, or their history is suggestive of REM sleep behavior disorder and an established synucleinopathy diagnosis like Parkinson's disease.

## Restless Leg Syndrome

It is noted that this disorder does not belong to the parasomnias, but it is another diagnosis in the sleep-wake disorders chapter. Restless leg syndrome (G25.81) is diagnosed in people who have an urge to move their legs, usually in response to uncomfortable and unpleasant sensations.

All of the following must be present for diagnosis:
- An urge to move legs, which begins or worsens during rest or inactivity.
- An urge to move legs which is partially or totally alleviated by movement.
- An urge to move legs that is worse in the evening or night when compared to day or that occurs only during the evening or night.

Other important criteria for diagnosis include:
- Symptoms occur at least 3 times per week and last for at least 3 months.
- Symptoms cause significant distress or impairment in functioning.
- Symptoms are not caused by another mental disorder or medical condition and are not attributable to the physiological effects of a substance.

## Substance/Medication-Induced Sleep Disorder

Substance/medication-induced sleep disorder is diagnosed in people who have

- A prominent, severe sleep disturbance with evidence that symptoms developed during or soon after a substance or medication intoxication, withdrawal, or exposure.
- The involved substance/medication is capable of creating a sleep disturbance.
- To be diagnosed with this disorder, another sleep disorder that is not substance or medication-induced cannot be the cause of the disturbance; for instance, if symptoms precede substance/medication use, this diagnosis is not appropriate.

Coding for this disorder is based on the type of substance and whether a person has a substance use disorder.

Specifiers include:

- Insomnia type
  - Daytime sleepiness type
  - Parasomnia type
  - Mixed type
- With onset during intoxication
- With onset during withdrawal
- With onset after medication use

## Other Specified Insomnia Disorder

Other specified insomnia disorder (G47.09) is diagnosed in people who have:

- Symptoms characteristic of insomnia disorder
- Symptoms do not meet the full criteria for insomnia disorder or another disorder in this chapter
- The clinician specifies the reason the person does not meet the full criteria for insomnia disorder or another disorder.

## Unspecified Insomnia Disorder

Unspecified insomnia disorder (G47.00) is diagnosed in people who have:

- Symptoms characteristic of insomnia disorder
- Symptoms do not meet the full criteria for insomnia disorder or another disorder in this chapter
- The clinician specifies the reason the person does not meet the full criteria for insomnia disorder or another disorder.

## Other Specified Hypersomnolence Disorder

Other specified hypersomnolence disorder (G47.19) is diagnosed in people who have:

- Symptoms characteristic of hypersomnolence disorder
- Symptoms do not meet the full criteria for hypersomnolence disorder or another disorder in this chapter
- The clinician specifies the reason the person does not meet the full criteria for hypersomnolence disorder or another disorder.

## Unspecified Hypersomnolence Disorder

Unspecified hypersomnolence disorder (G47.10) is diagnosed in people who have:

- Symptoms characteristic of hypersomnolence disorder, but they do not meet the full criteria for hypersomnolence disorder or another disorder in this chapter
- The clinician specifies the reason the person does not meet full criteria for hypersomnolence disorder or another disorder.

## Other Specified Sleep-Wake Disorder

Other specified sleep-wake disorder (F47.8) is diagnosed in people who have:

- Symptoms characteristic of a sleep-wake disorder
- Symptoms meet the full criteria for any other disorders in this chapter, nor do they qualify for diagnoses of other specified insomnia disorder or other specified hypersomnolence disorder.

- This diagnosis is made when the clinician specifies the reason a person does not meet full diagnostic criteria for another sleep-wake disorder.

## Unspecified Sleep-Wake Disorder

Unspecified sleep-wake disorder (G47.9) is diagnosed in people who have:

- Symptoms characteristic of a sleep-wake disorder
- Symptoms do not meet full criteria for any other disorders in this chapter, nor do they qualify for diagnoses of other specified insomnia disorder or other specified hypersomnolence disorder.
- This diagnosis is made when the clinician does not specify the reason a person does not meet the full diagnostic criteria for another sleep-wake disorder.

# CHAPTER 13

# Sexual Dysfunctions

The sexual dysfunctions chapter in the DSM-5-TR includes a diverse group of disorders that cause dysfunction in a person's ability to respond sexually or experience pleasure during sex. It Is possible for a person to have multiple sexual dysfunctions and, as such, be diagnosed with multiple disorders in this chapter.

The specific disorders in this chapter are detailed below.

## Delayed Ejaculation

Delayed ejaculation (F52.32) is diagnosed in people who show either of the following symptoms during almost all or all instances of partnered sexual interaction when they do not desire delay:

- Noticeable delay in ejaculation.
- Noticeably infrequent or absent ejaculation.

Other essential criteria for diagnosis include:

- The duration of symptoms is at least approximately 6 months.
- The person experiences clinically significant distress related to symptoms.
- Symptoms are not better explained by a nonsexual mental disorder, severe relationship distress or other stressors, or the effects of a substance/medication or another medical problem.

Specifiers include:

- Lifelong
- Acquired
- Generalized
- Situational
- Mild
- Moderate
- Severe

## Erectile Disorder

Erectile disorder (F52.21) is diagnosed in people who show:

- One or more of the following symptoms during almost all or all instances of partnered sexual interaction:
  - Noticeable difficulty obtaining an erection.
  - Noticeable difficulty maintaining an erection until sexual activity is complete.
  - Noticeably reduced erectile rigidity.

Other essential criteria for diagnosis include:

- The duration of symptoms is at least approximately 6 months.
- The person experiences clinically significant distress related to symptoms.
- A nonsexual mental disorder, severe relationship distress or other stressors, or the effects of a substance/medication or another medical problem do not better explain symptoms.

Specifiers include:

- Lifelong
- Acquired
- Generalized
- Situational
- Mild
- Moderate
- Severe

## Female Orgasmic Disorder

Female orgasmic disorder (F52.31) is diagnosed in people who show:

- Either of the following symptoms during almost all or all sexual interactions:
  - Noticeable delay in, infrequency of, or absence of orgasm.
  - Noticeably reduced intensity of orgasms.

Other important criteria for diagnosis include:

- The duration of symptoms is at least approximately 6 months.
- The person experiences clinically significant distress related to symptoms.
- A nonsexual mental disorder, severe relationship distress or other stressors, or the effects of a substance or another medical problem do not better explain symptoms.

Specifiers include:

- Lifelong
- Acquired
- Generalized
- Situational
- Mild
- Moderate
- Severe

## Female Sexual Interest/Arousal Disorder

Female sexual interest/arousal disorder (F52.22) is diagnosed in people who have:

- A lack of or a considerably reduced level of arousal or interest in sex, as evidenced by at least 3 of the following symptoms:
  - Absent or reduced interest in sex.
  - Absent or reduced thoughts or fantasies of a sexual or erotic nature.
  - Lack of or reduced initiation of sexual activity and tendency to be unreceptive to partner's initiation of sexual activity.

- Lack of or reduced sexual excitement or pleasure when engaged in sexual activity on almost all or all occasions.
- Lack of or reduced interest or arousal in reaction to sexual or erotic cues, which may be written, verbal, or visual.
- Lack of or reduced genital or non-genital sensations when engaged in sexual activity in almost all or all sexual encounters.

Other important criteria for diagnosis include:

- The duration of symptoms is at least 6 months.
- The person experiences clinically significant distress related to symptoms.
- A nonsexual mental disorder, severe relationship distress or other stressors, or the effects of a substance or another medical problem do not better explain symptoms.

Specifiers include:

- Lifelong
- Acquired
- Generalized
- Situational
- Mild
- Moderate
- Severe

## Genito-Pelvic Pain/Penetration Disorder

Genito-pelvic pain/penetration disorder (F52.6) is diagnosed in people who struggle with at least one of the following persistently:

- Vaginal penetration with intercourse.
- Significant vulvovaginal or pelvic pain with intercourse or attempts at penetration.
- Significant fear or anxiety related to vulvovaginal or pelvic pain when anticipating penetration, during penetration, or as a result of penetration.
- Significant tension or tightening in pelvic floor muscles when attempting vaginal penetration.

Other important criteria for diagnosis include:

- The duration of symptoms is at least 6 months.
- The person experiences clinically significant distress related to symptoms.
- A nonsexual mental disorder, severe relationship distress or other stressors, or the effects of a substance or another medical problem do not better explain symptoms.

Specifiers include:

- Lifelong
- Acquired
- Generalized
- Situational
- Mild
- Moderate
- Severe

## Male Hypoactive Sexual Desire Disorder

Male hypoactive sexual desire disorder (F52.0) is diagnosed in people with:

- Persistently or repeatedly show deficient or absent thoughts of a sexual or erotic nature.
  - Whether thoughts are deficient depends upon the clinician's judgment, who must consider factors influencing sexual functioning, such as age and sociocultural contexts.

Other important criteria for diagnosis include:

- The duration of symptoms is at least 6 months.
- The person experiences clinically significant distress related to symptoms.
- A nonsexual mental disorder, severe relationship distress or other stressors, or the effects of a substance or another medical problem do not better explain symptoms.

Specifiers include:

- Lifelong
- Acquired
- Generalized
- Situational
- Mild
- Moderate
- Severe

## Premature (Early) Ejaculation

Premature (early) ejaculation (F52.4) is diagnosed in people who show:

- An ongoing pattern of ejaculating within about 1 minute of vaginal penetration during partnered sexual activity
- The individual does not desire this early ejaculation, and at this time, criteria for premature (early) ejaculation during non-vaginal sexual activities have not been established.

Additional important criteria for diagnosis include:

- Symptoms of premature ejaculation must endure for at least 6 months and must be present in almost all or all sexual interactions.
- The person experiences clinically significant distress related to symptoms.
- A nonsexual mental disorder, severe relationship distress or other stressors, or the effects of a substance or another medical problem do not better explain symptoms.

Specifiers include:

- Lifelong
- Acquired
- Generalized
- Situational
- Mild
- Moderate
- Severe

## Substance/Medication-Induced Sexual Dysfunction

Substance/medication-induced sexual dysfunction is diagnosed in people who show:

- Clinically significant disruption in sexual functioning
- There is evidence that symptoms developed during or soon after a substance or medication intoxication, withdrawal, or exposure, and the involved substance/ medication can produce such symptoms.
- To be diagnosed with this disorder, the symptoms cannot be better explained by a sexual dysfunction that is not substance or medication-induced.

Coding for this disorder is based on the type of substance and whether a person has a substance use disorder.

Specifiers include:

- With onset during intoxication
- With onset during withdrawal
- With onset after medication use
- Mild
- Moderate
- Severe

## Other Specified Sexual Dysfunction

Other specified sexual dysfunction (F52.8) is diagnosed in people who have:

- Symptoms characteristic of sexual dysfunction
- Symptoms do not meet the full criteria for another disorder in this chapter
- The clinician specifies the reason the person does not meet the full criteria for another sexual dysfunction.

## Unspecified Sexual Dysfunction

Unspecified sexual dysfunction (F52.9) is diagnosed in people who have:

- Symptoms characteristic of sexual dysfunction
- Symptoms do not meet the full criteria for another disorder in this chapter
- The clinician does not specify the reason the person does not meet the full criteria for another sexual dysfunction.

# CHAPTER 14

# Gender Dysphoria

The gender dysphoria chapter in the DSM-5-TR includes just one overarching diagnosis of gender dysphoria, along with other specified and unspecified diagnoses of this disorder. Gender dysphoria is described below in more detail.

## Gender Dysphoria

Gender dysphoria (F64.2) is diagnosed in children who live with a noticeable incongruence between their experienced or expressed gender and the gender they were assigned at birth.

- To be diagnosed, a person must experience at least 6 of the following symptoms for at least 6 months, and at least one must be the first item on the list:
  - Desiring strongly to be the opposite gender or insisting that one is the opposite gender or some alternative gender aside from what was assigned biologically at birth.
  - In those assigned male at birth, a strong preference for cross-dressing or wearing female attire; in those assigned female at birth, a strong preference for wearing masculine clothing and a resistance to wearing feminine clothing.
  - Strongly preferring cross-gender roles when engaged in make-believe or fantasy play.
  - Strongly preferring toys, games, or activities typically associated with the opposite gender.
  - Strongly preferring playmates of the opposite gender.

- In those assigned male at birth, firmly rejecting typically masculine toys, games, and activities while avoiding rough and tumble play; in those assigned female at birth, strongly rejecting typically feminine toys, games, and activities.
- Strongly disliking one's sexual anatomy.
- Strongly desiring primary or secondary sex characteristics that match one's identified gender.

Gender dysphoria (F64.0) is diagnosed in adolescents and adults who live with a noticeable incongruence between their experienced or expressed gender and the gender they were assigned at birth.

- To be diagnosed, a person must experience at least 2 of the following symptoms for at least 6 months:
  - Significant incongruence between one's experienced or expressed gender and primary or secondary sex characteristics, which may involve an incongruence with anticipated secondary sex characteristics in adolescents.
  - Strongly desiring to rid oneself of primary or secondary sex characteristics due to incongruence with one's experienced or expressed gender in adolescents may present as wanting to prevent the development of secondary sex characteristics.
  - Strongly desiring primary and/or secondary sex characteristics of the opposite gender.
  - Strongly desiring to be of the opposite gender or another alternative gender different from what one was assigned at birth.
  - Strongly desiring to be treated as the opposite gender or another alternative gender different from what one was assigned at birth.
  - A strong feeling that one's feelings and reactions are typical of the opposite gender or some alternative gender different from one's assigned gender.

Specifiers include:

- With a disorder/difference of sex development
- Post transition

## Other Specified Gender Dysphoria

Other specified gender dysphoria (F64.8) is diagnosed in people who have:

- Symptoms of gender dysphoria that do not meet the full criteria for this disorder
- The clinician specifies the reason the person does not meet the full criteria for gender dysphoria.

## Unspecified Gender Dysphoria

Unspecified gender dysphoria (F64.9) is diagnosed in people who have:

- Symptoms of gender dysphoria
- Symptoms do not meet the full criteria for this disorder
- The clinician does not specify the reason the person does not meet the full criteria for gender dysphoria.

# CHAPTER 15

# Disruptive, Impulse-Control, and Conduct Disorders

The disruptive, impulse-control, and conduct disorders chapter in the DSM-5-TR includes various mental health conditions that are characterized by a lack of control over one's emotions and behaviors. What separates these disorders from other disorders involving difficulties with emotional and behavioral regulation is that disorders in the disruptive, impulse-control, and conduct disorders chapter involve behaviors that violate the rights of other people.

The specific disorders in this chapter are described in more detail below.

## Oppositional Defiant Disorder

Individuals with oppositional defiant disorder (F91.3) show:

- An angry or irritable mood, argumentative/defiant behavior, or vindictiveness that persists for at least 6 months and causes at least 4 symptoms from any of the following categories; to be diagnosed, these behaviors must occur when interacting with at least one person who isn't a sibling:
  - Angry/Irritable Mood Symptoms:
    - Losing temper often.
    - Being touchy or easily annoyed.
    - Being angry and resentful often.
  - Argumentative/Defiant Behavior Symptoms:
    - Arguing often with authority figures or adults if one is a child/adolescent.

- Defying authority often or refusing to comply with the requests of authority figures.
- Frequently deliberately annoying others.
- Frequently blaming others for one's mistakes or misbehavior.
  - Vindictiveness Symptoms
    - Being spiteful or vindictive at least once in the previous 6 months.

Other important criteria for diagnosis include:

- Behaviors occur most days for at least 6 months in children under 5 years old.
- Behaviors occur at least once weekly for at least 6 months in children 5 years or older.
- Behavior causes distress in the person diagnosed or in the person's immediate social context, which includes family, peers, and colleagues; alternatively, the behavior negatively impacts functioning in social, educational, or occupational settings.

Specifiers include:

- Mild
- Moderate
- Severe

## Intermittent Explosive Disorder

Intermittent explosive disorder (F63.81) is diagnosed in people who show:

- Repeated behavioral outbursts that involve a failure to control aggression.
- The disorder can be manifested by either one of the following:
  - Outbursts of verbal aggression, which may involve temper tantrums, tirades, verbal arguments, or physical aggression toward property, animals, or other individuals,
  - Outbursts occur an average of twice weekly for 3 months
  - If physical aggression, it does not cause damage or destruction to property or bodily injury.
- Over 12 months, 3 outbursts that cause damage or destruction to property or involve physical injury against animals or others.

Other important criteria for diagnosis include:

- The level of aggression displayed during the outbursts is significantly disproportionate to the provocation or any stressors precipitating it.
- The behavioral outbursts are not premeditated, so they are impulsive or anger-based and are not carried out to achieve a tangible objective like money or power.
- The behavioral outbursts result in significant distress in the person diagnosed, impairment in occupational or social functioning, or financial or legal consequences.
- The person with the diagnosis is at least 6 years of age.

## Conduct Disorder

Conduct disorder is diagnosed in people who show:

- An ongoing pattern of behavior that involves violating others' fundamental rights or violating social norms or rules.
- At least 3 of 15 symptoms from any of the following categories over the previous 12 months, and at least one symptom must be present within the prior 6 months:
  - Symptoms of Aggression to People and Animals:
    - Bullying, threatening, or intimidating others frequently.
    - Initiating physical fights frequently.
    - History of using a weapon, like a bat or a knife that can cause serious physical harm to other people.
    - History of physical cruelty to people.
    - History of physical cruelty to animals.
    - History of stealing after confronting a victim, such as would happen with a mugging.
    - History of forcing someone into sexual activity.
  - Symptoms of Destruction of Property:
    - History of deliberately setting fires with the intention to cause severe damage.
    - History of intentionally destroying others' property by methods other than fire setting.

- ○ Symptoms of Deceitfulness or Theft:
  - History of breaking into someone's home, building, or car.
  - Frequently lying to obtain items or favors or to avoid obligations.
  - History of stealing valuable items without confronting the victim.
- ○ Symptoms of Serious Violations of Rules:
  - Frequently staying out at night against parents' wishes, before age 13.
  - History of running away from home overnight at least 2 times while living with parents or guardians or at least once without returning home for a lengthy period.
  - Frequent truancy from school before age 13.

Other important criteria for diagnosis include:

- Behavioral disturbance causes clinically significant problems with academic, social, or occupational functioning.
- If the person is at least 18 years of age, they do not meet the criteria for antisocial personality disorder.

Specifiers include:

- Childhood-onset type: coded as F91.1
- Adolescent-onset type: coded as F91.2
- Unspecified onset: coded as F91.9
- With limited prosocial emotions
- Mild
- Moderate
- Severe

## Antisocial Personality Disorder

It is worth noting that antisocial personality disorder is listed due to its close relationship to externalizing conduct disorders in this chapter. However, the full criteria for antisocial personality disorder are listed in the Personality Disorders chapter.

## Pyromania

Pyromania (F63.1) is diagnosed in people who deliberately set fires more than once. Other diagnostic criteria include:

- Tension or emotional arousal precedes fire setting.
- Showing fascination with, interest in, curiosity about, or attraction to fire and factors like paraphernalia and consequences of fire.
- Experiencing pleasure, gratification, or relief with or in the aftermath of fire setting.
- Fire setting is not completed for financial gain, to express sociopolitical ideology, to hide criminal activity, to express anger or vengeance, to improve living circumstances, because of a hallucination or delusion, or because of impaired judgment.
- Conduct disorder, manic episodes, and antisocial personality disorder do not better account for the fire setting.

## Kleptomania

Kleptomania (F63.2) is diagnosed in people who:repeatedly fail to resist the urge to steal items they don't need or that don't offer significant monetary value. Other diagnostic criteria include:

- Increase in tension immediately before committing theft.
- Experiencing pleasure, gratification, or relief when committing theft.
- The theft doesn't occur to express anger or vengeance and is not the result of a delusion or hallucination.
- Conduct disorder, manic episodes, and antisocial personality disorder do not better account for the stealing.

## Other Specified Disruptive, Impulse Control, and Conduct Disorder

Other specified disruptive, impulse control, and conduct disorder (F91.8) is diagnosed in people who show:

- Symptoms characteristic of a disorder in this chapter
- Symptoms do not meet full diagnostic criteria for another disorder
- The clinician specifies the reason full diagnostic criteria for another disorder are not met.

## Unspecified Disruptive, Impulse Control, and Conduct Disorder

Unspecified disruptive, impulse control, and conduct disorder (F91.9) is diagnosed in people who show:

- Symptoms characteristic of a disorder in this chapter
- Symptoms do not meet full diagnostic criteria for another disorder
- The clinician does not specify the reason full diagnostic criteria for another disorder are not met.

# CHAPTER 16

# Substance-Related and Addictive Disorders

The substance-related and addictive disorders chapter of the DSM-5-TR includes addictions or substance use disorders related to multiple different classes of drugs and other substances, such as caffeine. Rather than referring to these disorders as "addictions," they are classified as substance use disorders. There is also one non-substance-related disorder in this chapter: gambling disorder.

The specific disorders in this chapter are described in more detail below. Given that numerous substances may be associated with a substance use disorder, this chapter begins with general criteria for a substance use disorder and then provides coding based on the specific substance. Then, the chapter proceeds to describe disorders involving intoxication or withdrawal.

## Substance Use Disorders

The general criteria for a substance use disorder are as follows; keep in mind that the exact name of the disorder depends upon the substance of misuse*:

- A problematic pattern of use that leads to significant impairment or distress, which involves at least 2 of the following symptoms within 12 months:
    - Using larger amounts of the substance or using for a longer time than intended.
    - Being unable to reduce or control the use of the substance, even if one desires to do so.

- Spending a significant amount of time acquiring or using the substance or recovering from the effects of use.
- Experiencing cravings for the substance.
- Using the substance to the extent that one is unable to fulfill duties at work, school, or home.
- Continuing to use the substance, even when it causes or worsens ongoing or repeated social or personal problems.
- Giving up social, work-related, or recreational activities due to substance misuse.
- Ongoing substance misuse in situations that cause physical hazards.
- Continuing substance misuse even when one knows it is causing or worsening a physical or psychological problem.
- Experiencing tolerance can manifest as increasing amounts of the substance to achieve the same effect or as a reduced effect with ongoing use of the same amount
- Withdrawal, which can manifest as experiencing a characteristic withdrawal symptom associated with the substance or using the substance to avoid or relieve withdrawal side effects.

*The above criteria are applied to all substance use disorders, aside from phencyclidine use disorder, other hallucinogen use disorder, and inhalant use disorder. Diagnostic criteria for these disorders include the first 10 symptoms listed above, but withdrawal symptoms are not established for these substances, so the withdrawal criterion does not apply.

Below, the specifiers and related codes for each of the substance use disorders are detailed based on the specific substance of misuse.

## Alcohol Use Disorder

- In early remission
- In sustained remission
- In a controlled environment
- Mild (diagnosed when 2-3 symptoms are present): F10.10
- Mild, in early remission: F10.11
- Mild, in sustained remission: F10.11
- Moderate (diagnosed when 4-5 symptoms are present): F10.20

- Moderate, in early remission: F10.21
- Moderate, in sustained remission: F10.21
- Severe (diagnosed when 6 or more symptoms are present): F10.20
- Severe, in early remission: F10.21
- Severe, in sustained remission: F10.21

## Cannabis Use Disorder

- In early remission
- In sustained remission
- In a controlled environment
- Mild (diagnosed when 2-3 symptoms are present): F12.10
- Mild, in early remission: F12.11
- Mild, in sustained remission F12.11
- Moderate (diagnosed when 4-5 symptoms are present): F12.20
- Moderate, in early remission: F12.21
- Moderate, in sustained remission: F12.21
- Severe (diagnosed when 6 or more symptoms are present): F12.20
- Severe, in early remission: F12.21
- Severe, in sustained remission: F12.21

## Phencyclidine Use Disorder, Other Hallucinogen Use Disorder

- In early remission
- In sustained remission
- In a controlled environment
- Mild (diagnosed when 2-3 symptoms are present): F16.10
- Mild, in early remission: F16.11
- Mild, in sustained remission: F16.11
- Moderate (diagnosed when 4-5 symptoms are present): F16.20
- Moderate, in early remission: F16.21
- Moderate, in sustained remission: F16.21
- Severe (diagnosed when 6 or more symptoms are present): F16.20
- Severe, in early remission: F16.21
- Severe, in sustained remission: F16.21

## Inhalant Use Disorder

- In early remission
- In sustained remission
- In a controlled environment
- Mild (diagnosed when 2-3 symptoms are present): F18.10
- Mild, in early remission: F18.11
- Mild, in sustained remission: F18.11
- Moderate (diagnosed when 4-5 symptoms are present): F18.20
- Moderate, in early remission: F18.21
- Moderate, in sustained remission: F18.21
- Severe (diagnosed when 6 or more symptoms are present): F18.20
- Severe, in early remission: F18.21
- Severe, in sustained remission: F18.21

## Opioid Use Disorder

- In early remission
- In sustained remission
- In a controlled environment
- Mild (diagnosed when 2-3 symptoms are present): F11.10
- Mild, in early remission: F11.11
- Mild, in sustained remission: F11.11
- Moderate (diagnosed when 4-5 symptoms are present): F11.20
- Moderate, in early remission: F11.21
- Moderate, in sustained remission: F11.21
- Severe (diagnosed when 6 or more symptoms are present): F11.20
- Severe, in early remission: F11.21
- Severe, in sustained remission: F11.21

## Sedative, Hypnotic, and Anxiolytic-Related Disorders

- In early remission
- In sustained remission
- In a controlled environment
- Mild (diagnosed when 2-3 symptoms are present): F13.10
- Mild, in early remission: F13.11

- Mild, in sustained remission: F13.11
- Moderate (diagnosed when 4-5 symptoms are present): F13.20
- Moderate, in early remission: F13.21
- Moderate, in sustained remission: F13.21
- Severe (diagnosed when 6 or more symptoms are present): 13.20
- Severe, in early remission: F13.21
- Severe, in sustained remission: F13.21

## Stimulant Use Disorder

- In early remission
- In sustained remission
- In a controlled environment
- Mild (diagnosed when 2-3 symptoms are present): F15.10 (amphetamine-type substance; F14.10 (cocaine); F15.10 (other or unspecified stimulant)
- Mild, in early remission: F15.11 (amphetamine-type substance; F14.11 (cocaine); F15.11 (other or unspecified stimulant)
- Mild, in sustained remission: F15.11 (amphetamine-type substance; F14.11 (cocaine); F15.11 (other or unspecified stimulant)
- Moderate (diagnosed when 4-5 symptoms are present): F15.20 (amphetamine-type substance); F14.20 (cocaine); F15.20 (other or unspecified stimulant)
- Moderate, in early remission: F15.21 (amphetamine-type substance); F14.21 (cocaine); F15.21 (other or unspecified stimulant)
- Moderate, in sustained remission: F15.21 (amphetamine-type substance); F14.21 (cocaine); F15.21 (other or unspecified stimulant)
- Severe (diagnosed when 6 or more symptoms are present): F15.20 (amphetamine-type substance); F14.20 (cocaine); F15.20 (other or unspecified stimulant)
- Severe, in early remission: F15.21 (amphetamine-type substance); F14.21 (cocaine); F15.21 (other or unspecified stimulant)
- Severe, in sustained remission: F15.21 (amphetamine-type substance); F14.21 (cocaine); F15.21 (other or unspecified stimulant)

## Tobacco Use Disorder

- In early remission
- In sustained remission
- On maintenance therapy
- In a controlled environment
- Mild (diagnosed when 2-3 symptoms are present): Z72.0
- Moderate (diagnosed when 4-5 symptoms are present): F17.200
- Moderate, in early remission: F17.201
- Moderate, in sustained remission: F17.201
- Severe (diagnosed when 6 or more symptoms are present): F17.200
- Severe, in early remission: F17.201
- Severe, in sustained remission: F17.201

## Other (or Unknown) Substance Use Disorder

- In early remission
- In sustained remission
- In a controlled environment
- Mild (diagnosed when 2-3 symptoms are present): F19.10
- Mild, in early remission: F19.11
- Mild, in sustained remission: F19.11
- Moderate (diagnosed when 4-5 symptoms are present): F19.20
- Moderate, in early remission: F19.21
- Moderate, in sustained remission: F19.21
- Severe (diagnosed when 6 or more symptoms are present): F19.20
- Severe, in early remission: F19.21
- Severe, in sustained remission: F19.21

## Other Disorders in the Chapter

The substance-related and addictive disorders chapter in the DSM-5-TR is quite lengthy, including numerous disorders. The additional disorders in this chapter are described in detail below.

## Alcohol Intoxication

An alcohol intoxication diagnosis is given in people who have recently ingested alcohol and who show problematic behavior or psychological changes during or soon after ingestion.

- To be diagnosed, a person must show at least one of the following symptoms related to alcohol use:
    - Slurred speech.
    - Lack of coordination.
    - Unsteady gait.
    - Nystagmus.
    - Difficulties with attention or memory.
    - Stupor or coma.

Coding is as follows:

- Comorbid mild alcohol use disorder: F10.120
- Comorbid moderate or severe alcohol use disorder: F10.220
- No comorbid alcohol use disorder: F10.920

## Alcohol Withdrawal

Alcohol withdrawal is diagnosed in people who stop or reduce heavy, prolonged alcohol use.

- To be diagnosed, a person must show at least 2 of the following symptoms, beginning a few hours to a few days after stopping or reducing alcohol use:
    - Hyperactivity in the autonomic nervous system manifests in symptoms like sweating or high pulse rate.
    - Hand tremor.
    - Insomnia.
    - Nausea or vomiting.
    - Hallucinations which are transient and may be visual, tactile, or auditory.
    - Psychomotor agitation.
    - Anxiety.
    - Generalized tonic-clonic seizures.

Specifiers and coding are as follows:

- With perceptual disturbances
- Without perceptual disturbances, with mild alcohol use disorder comorbid: F10.130
- Without perceptual disturbances, with moderate or severe alcohol use disorder comorbid: F10.230
- Without perceptual disturbances, with no comorbid alcohol use disorder: F10.930
- With perceptual disturbances, with mild alcohol use disorder comorbid: F10.132
- With perceptual disturbances, with moderate or severe alcohol use disorder comorbid: F10.232
- With perceptual disturbances, with no comorbid alcohol use disorder: F10.932

## Caffeine Intoxication

Caffeine intoxication (F15.920) is diagnosed in people who have recently consumed caffeine, usually in high doses considerably above 250 milligrams, and who show at least 5 of the following symptoms shortly after caffeine use:

- Restlessness.
- Nervousness.
- Excitement.
- Insomnia.
- Flushed face.
- Diuresis.
- Gastrointestinal disturbance.
- Muscle twitching.
- Rambling flow of thoughts and speech.
- Tachycardia or cardiac arrhythmia.
- Periods of inexhaustibility.
- Psychomotor agitation.

## Caffeine Withdrawal

Caffeine withdrawal (F15.93) is diagnosed in people who show prolonged daily use of caffeine and who suddenly stop or reduce caffeine use.

- To be diagnosed, a person must show at least 3 of the following symptoms within 24 hours of stopping or reducing caffeine use:
    - Headache.
    - Significant fatigue or drowsiness.
    - Dysphoric or depressed mood or irritability.
    - Concentration problems.
    - Flu-like symptoms including nausea, vomiting, or muscle pain and stiffness.

## Cannabis Intoxication

Cannabis intoxication is diagnosed in people who have recently used cannabis and who show problematic behavior or psychological changes during or soon after use.

- To be diagnosed, a person must show at least 2 of the following symptoms related to cannabis use, and symptoms must present within 2 hours of use:
    - Conjunctival injection.
    - Increase in appetite.
    - Dry mouth.
    - Tachycardia.

Specifiers and coding are as follows:
- With perceptual disturbances.
- Without perceptual disturbances, with mild cannabis use disorder comorbid: F12.120
- Without perceptual disturbances, with moderate or severe cannabis use disorder comorbid: F12.220
- Without perceptual disturbances, with no comorbid alcohol use disorder: F12.920
- With perceptual disturbances, with mild cannabis use disorder comorbid: F12.122

- With perceptual disturbances, with moderate or severe cannabis use comorbid: F12.222
- With perceptual disturbances, with no comorbid cannabis use disorder: F12.922

## Cannabis Withdrawal

Cannabis withdrawal is diagnosed in people who stop heavy, prolonged cannabis use and experience at least 3 of the following symptoms within a week after stopping use:

- Irritability, anger, or aggressive behavior.
- Nervousness or anxiety.
- Difficulty sleeping.
- Reduced appetite or weight loss.
- Restlessness.
- Depressed mood.
- Significantly uncomfortable physical symptoms, which may include abdominal pain, shakiness or tremors, sweating, fever, chills, or headache.

Coding is as follows:

- With mild cannabis use disorder comorbid: F12.13
- Moderate or severe cannabis use disorder comorbid: F12.23
- No cannabis use disorder comorbid: F12.93

## Phencyclidine Intoxication

Phencyclidine intoxication is diagnosed in people who have recently used phencyclidine or a similar substance and who develop problematic behavioral changes during or shortly after the use of the substance.

- To be diagnosed, a person must show at least 2 of the following symptoms within an hour of phencyclidine use:
  - Vertical or horizontal nystagmus.
  - Hypertension or tachycardia
  - Numbness or reduced responsiveness to pain.

- Ataxia.
- Dysarthria.
- Rigidity in muscles.
- Seizures or coma.
- Hyperacusis.

Coding is as follows:

- With mild comorbid phencyclidine use disorder: F16.120
- With moderate or severe phencyclidine use disorder: F16.220
- No phencyclidine use disorder comorbid: F16.920

## Other Hallucinogen Intoxication

Other hallucinogen intoxication is diagnosed in people who have recently used a hallucinogen other than phencyclidine and who develop problematic behavioral or psychological changes during or shortly after use.

- To be diagnosed, a person must develop perceptual changes while fully awake and alert, including intensifying perceptions, depersonalization, derealization, illusions, or hallucinations.
- A person must also show at least 2 of the following symptoms for diagnosis:
  - Dilated pupils.
  - Tachycardia.
  - Sweating.
  - Palpitations.
  - Blurred vision.
  - Tremors.
  - Incoordination.

Coding is as follows:

- With mild comorbid hallucinogen use disorder: F16.120
- With moderate or severe hallucinogen use disorder: F16.220
- No hallucinogen use disorder comorbid: F16.920

## Hallucinogen Persisting Perception Disorder

Hallucinogen persisting perception disorder (F16.983) is diagnosed in people who have stopped using a hallucinogen and who re-experience at least one of the perceptual symptoms they experienced while intoxicated with the hallucinogen. Symptoms might include:

- Geometric hallucinations
- Perceiving false movements in one's peripheral visual field
- Intensified colors
- Trails of images of objects in motion
- Positive afterimages
- Halos around objects
- Macropsia/micropsia.

## Inhalant Intoxication

Inhalant intoxication is diagnosed in people who have recently been exposed to high doses of inhalant substances, whether intended or unintended. Exposure might include substances like volatile hydrocarbons like toluene or gasoline. To be diagnosed, a person must experience significant problematic behavior or psychological changes during or shortly after inhalant exposure, and they must show at least 2 of the following symptoms:

- Dizziness.
- Nystagmus.
- Lack of coordination.
- Slurred speech.
- Unsteady gait.
- Lethargy.
- Depressed reflexes.
- Psychomotor retardation.
- Tremor.
- Generalized weakness in the muscles.
- Blurred vision or diplopia.
- Stupor or coma.
- Euphoria.

Coding is as follows:

- With mild inhalant use disorder comorbid: F18.120
- With moderate or severe inhalant use disorder comorbid: F18.220
- No inhalant use disorder comorbid: F18.920

## Opioid Intoxication

Opioid intoxication is diagnosed in people who have recently used an opioid and who show significant problematic behavioral or psychological changes during or shortly after opioid use. To be diagnosed, a person must experience pupillary constriction or dilation as well as at least one of the following symptoms, which develop during or shortly after opioid use:

- Drowsiness or coma.
- Slurred speech.
- Impaired attention or memory.

Specifiers and coding are as follows:

- With perceptual disturbances
- Without perceptual disturbances, with mild opioid use disorder comorbid: F11.120
- Without perceptual disturbances, with moderate or severe opioid use disorder comorbid: F11.220
- Without perceptual disturbances, with no opioid use disorder comorbid: F11.920
- With perceptual disturbances, with mild opioid use disorder comorbid: F11.122
- With perceptual disturbances, with moderate or severe opioid use disorder comorbid: F11.222
- With perceptual disturbances, with no opioid use disorder comorbid: F11.922

## Opioid Withdrawal

Opioid withdrawal is diagnosed in people who stop or reduce opioid use after a period of prolonged, heavy use. Alternatively, the diagnosis may be given to people who are using an opioid antagonist after a period of opioid use. To be diagnosed, a person must also experience at least 3 of the following symptoms:

- Dysphoric mood.
- Nausea or vomiting.
- Muscle aches.
- Lacrimation or rhinorrhea.
- Dilated pupils, piloerection, or sweating.
- Diarrhea.
- Yawning.
- Fever.
- Insomnia.

Specifiers and coding are as follows:

- With mild opioid use disorder comorbid: F11.13
- With moderate or severe opioid use disorder comorbid: F11.23
- With no opioid use disorder comorbid: F11.93

## Sedative, Hypnotic, or Anxiolytic Intoxication

Sedative, hypnotic, or anxiolytic intoxication is diagnosed in people who have recently used a sedative, hypnotic, or anxiolytic and who show significant behavioral or psychological changes during or shortly after use. To be diagnosed, a person must also show at least one of the following symptoms:

- Slurred speech.
- Lack of coordination.
- Unsteady gait.
- Nystagmus.
- Impaired cognition, such as difficulties with attention or memory.
- Stupor or coma.

Specifiers and coding are as follows:

- With mild sedative, hypnotic, or anxiolytic use disorder comorbid: F13.120
- With moderate or severe sedative, hypnotic, or anxiolytic use disorder comorbid: F13.220
- With no comorbid sedative, hypnotic, or anxiolytic use disorder present: F13.920

## Sedative, Hypnotic, or Anxiolytic Withdrawal

Sedative, hypnotic, or anxiolytic withdrawal is diagnosed in people who stop or reduce prolonged use of a sedative, hypnotic, or anxiolytic and who show at least 2 of the following symptoms, beginning within several hours to a few days after stopping use of the substance:

- Evidence of the autonomic nervous system's hyperactivity like sweating or high pulse rate.
- Hand tremor.
- Insomnia.
- Nausea or vomiting.
- Transient hallucinations which may be visual, tactile, or auditory.
- Psychomotor agitation.
- Anxiety.
- Grand mal seizures.

Specifiers and coding are as follows:

- With perceptual disturbances
- Without perceptual disturbances, with mild sedative, hypnotic, or anxiolytic use disorder comorbid: F13.130
- Without perceptual disturbances, with moderate or severe sedative, hypnotic, or anxiolytic use disorder comorbid: F13.230
- Without perceptual disturbances, with no sedative, hypnotic, or anxiolytic use disorder comorbid: F13.930
- With perceptual disturbances, with mild sedative, hypnotic, or anxiolytic use disorder comorbid: F13.132

- With perceptual disturbances, with moderate or severe sedative, hypnotic, or anxiolytic use disorder comorbid: F13.232
- With perceptual disturbances, with no sedative, hypnotic, or anxiolytic use disorder comorbid: F13.932

## Stimulant Intoxication

Stimulant intoxication is diagnosed in people who have recently used an amphetamine-type substance, cocaine, or another stimulant and who show significant problematic behavioral or psychological changes during or shortly after stimulant use. To be diagnosed, a person must show:

- At least 2 of the following symptoms:
  - Tachycardia or bradycardia.
  - Dilated pupils.
  - Increased or lowered blood pressure.
  - Perspiration or chills.
  - Nausea or vomiting.
  - Evidence of weight loss.
  - Psychomotor agitation or retardation.
  - Weakness in the muscles, respiratory depression, chest pain, or cardiac arrhythmias.

Specifiers and coding are as follows:
- With perceptual disturbances.
- Without perceptual disturbances, with mild amphetamine-type substance or other stimulant use disorder comorbid: F15.120
- Without perceptual disturbances, with moderate or severe amphetamine- type substance or other stimulant use disorder comorbid: F15.220
- Without perceptual disturbances, with no comorbid amphetamine-type substance or other stimulant use disorder: F15.920
- Without perceptual disturbances, with mild cocaine use disorder comorbid: F14.120
- Without perceptual disturbances, with moderate or severe cocaine use disorder comorbid: F14.220

- Without perceptual disturbances, with no comorbid cocaine use disorder: F14.920
- With perceptual disturbances, with mild amphetamine-type substance or other stimulant use disorder comorbid: F15.122
- With perceptual disturbances, with moderate or severe amphetamine-type substance or other stimulant use disorder comorbid: F15.222
- With perceptual disturbances, with no comorbid amphetamine-type substance or other stimulant use disorder: F15.922
- With perceptual disturbances, with mild cocaine use disorder comorbid: F14.122
- With perceptual disturbances, with moderate to severe cocaine use disorder present: F14.222
- With perceptual disturbances, with no comorbid cocaine use disorder: F14.922

## Stimulant Withdrawal

Stimulant withdrawal is diagnosed in people who stop or reduce long-term use of an amphetamine-type substance, cocaine, or another stimulant and who show:

- Dysphoric mood and at least 2 of the following physiological changes a few hours to several days after stopping or reducing stimulant use:
  - Fatigue.
  - Vivid, unpleasant dreams.
  - Insomnia or hypersomnia.
  - Increase in appetite.
  - Psychomotor retardation or agitation.

Specifiers and coding are as follows:

- With mild amphetamine-type substance or other stimulant use disorder comorbid: F15.13
- With moderate or severe amphetamine-type substance or other stimulant use disorder comorbid: F15.23
- With no comorbid amphetamine-type substance or other stimulant use disorder: F15.93

- With mild cocaine use disorder comorbid: F14.13
- With moderate or severe cocaine use disorder comorbid: F14.23
- With no comorbid cocaine use disorder: F14.93

## Tobacco Withdrawal

Tobacco withdrawal (F17.203) is diagnosed in people who have used tobacco daily for at least several weeks and who suddenly stop their use. To be diagnosed, a person must show;

- At least 4 of the following symptoms within 24 hours of stopping:
    - Irritability, frustration, or anger.
    - Anxiety.
    - Difficulty concentrating.
    - Increased appetite.
    - Restlessness.
    - Depressed mood.
    - Insomnia.

## Other (or Unknown) Substance Intoxication

Other (or unknown) substance intoxication is diagnosed in people who experience:

- A reversible, substance-specific syndrome as a result of recent ingestion of or exposure to a substance not listed elsewhere in this chapter or that is unknown.
- The person must also experience clinically significant problem behaviors or psychological changes in response to substance use and its effects on the central nervous system.

Specifiers and coding are as follows:

- With perceptual disturbances
- Without perceptual disturbances with a mild other (or unknown) substance use disorder comorbid: F19.120
- Without perceptual disturbances with a moderate or severe other (or unknown) substance use disorder comorbid: F19.220

- Without perceptual disturbances, with no other (or unknown) substance use disorder comorbid: F19.920
- With perceptual disturbances, with mild other (or unknown) substance use disorder comorbid: F19.122
- With perceptual disturbances, with moderate or severe other (or unknown) substance use disorder present: F19.222
- With perceptual disturbances, with no comorbid other (or unknown) substance use disorder comorbid: F19.922

## Other (or Unknown) Substance Withdrawal

Other (or unknown) substance withdrawal is diagnosed in people who stop or reduce the use of a substance after heavy and prolonged use and who develop a substance-specific syndrome shortly after stopping or reducing their use.

This diagnosis is reserved for substances other than alcohol, caffeine, cannabis, opioids, sedatives, hypnotics, anxiolytics, stimulants, or tobacco. It can also be diagnosed when the symptoms of the substance causing it are unknown.

Specifiers and coding are as follows:

- With perceptual disturbances
- Without perceptual disturbances with a mild other (or unknown) substance use disorder comorbid: F19.130
- Without perceptual disturbances with a moderate or severe other (or unknown) substance use disorder comorbid: F19.230
- Without perceptual disturbances, with no other (or unknown) substance use disorder comorbid: F19.330
- With perceptual disturbances, with mild other (or unknown) substance use disorder comorbid: F19.132
- With perceptual disturbances, with moderate or severe other (or unknown) substance use disorder present: F19.232
- With perceptual disturbances, with no comorbid other (or unknown) substance use disorder comorbid: F19.932

## Unspecified Disorders

The unspecified disorders are diagnosed when a person shows:

- Symptoms characteristic of a substance use disorder
- Symptoms do not meet full criteria for a specific disorder in this chapter.

Coding is based on the type of substance and additional specifiers, as follows:

- Alcohol-related: F10.99
- Caffeine-related: F15.99
- Cannabis-related: F12.99
- Phencyclidine-related: F16.99
- Hallucinogen-related: F16.99
- Inhalant-related: F18.99
- Opioid-related: F11.99
- Sedative, hypnotic, or anxiolytic-related: F13.99
- Stimulant related: F15.99 (amphetamine-type substance or other stimulant); F14.99 (cocaine)
- Tobacco-related: F17.209
- Other (or unknown) substance: F19.99

## Gambling Disorder

Gambling disorder (F63.0) is not a substance-related disorder. It is diagnosed in people who show:

- Persistent, recurrent problematic behavior related to gambling.
- To be diagnosed, a person must show at least 4 of the following symptoms over 12 months:
  - Needing to gamble with increasing amounts of money to achieve the desired excitement from the activity.
  - Becoming restless or irritable when making attempts to reduce or stop gambling.
  - Showing a preoccupation with gambling.
  - Gambling often when under distress.
  - Returning another day to "get even" after losing money gambling.

- Lying to hide the extent of one's gambling.
- Jeopardizing or losing a significant relationship, job, educational, or career opportunity as a result of gambling.
- Relying on other people to provide funds to correct financial harms arising from gambling.
- The gambling behavior cannot be explained by a manic episode.

Specifiers include:

- Episodic
- Persistent
- In early remission
- In sustained remission
- Mild: When 4-5 criteria are met
- Moderate: When 6-7 criteria are met
- Severe: When 8-9 criteria are met

# CHAPTER 17

# Neurocognitive Disorders

The neurocognitive disorders chapter in the DSM-5-TR includes various disorders that involve deficits in cognitive functioning. In neurocognitive disorders, these deficits are acquired and not developmental, so they are absent from birth or early life and represent a decline in cognitive functioning. The chapter includes delirium as well as major and mild neurocognitive disorders and various subtypes of these disorders. It is worth noting that the term dementia may be commonly used in medical settings, primarily when referring to older adults living with degenerative dementias, but the proper term is a neurocognitive disorder.

The specific disorders in this chapter are detailed below.

## Delirium

Delirium is diagnosed in people who show:

- A disturbance in attention and a reduced awareness of the environment, which develops quickly, usually over several hours to a few days.
- This disturbance represents a change from baseline levels of attention and awareness and usually fluctuates in severity over a day.
- To be diagnosed, a person must also show one additional disturbance in cognition, such as a memory deficit, disorientation, language problems, perceptual issues, or declines in visuospatial ability.

Specifiers include:

- Acute
- Persistent
- Hyperactive
- Hypoactive
- Mixed level of activity

Coding is based on the following specifiers:

- Substance intoxication delirium (with specific code based upon the substance and whether a person has a use disorder)
- Substance withdrawal delirium (with the specific code based on the substance and whether a person has a use disorder)
- Medication-induced delirium
- Delirium due to another medical condition (F05)
- Delirium due to multiple etiologies (F05)

## Other Specified Delirium

Other specified delirium (R41.0) is diagnosed in people who show:

- Symptoms of delirium that do not meet the full criteria for delirium
- The clinician specifies the reason the full delirium criteria are not met.

## Unspecified Delirium

Unspecified delirium (R41.0) is diagnosed in people who show symptoms of delirium:

- Symptoms do not meet the full criteria for delirium
- The clinician does not specify why the full delirium criteria are not met.

## The Major and Mild Neurocognitive Disorders

The following section in this chapter details the major and mild neurocognitive disorders, beginning with the primary diagnostic criteria and then continuing with descriptions of the specific subtypes of these disorders.

## Major Neurocognitive Disorder

Major neurocognitive disorder is diagnosed in people who show:

- A significant cognitive decline from their previous level of functioning in at least one domain of cognitive functioning, which could include complex attention, executive function, learning and memory, language, perceptual-motor, or social cognition.
- Diagnosis is based on the following 2 criteria:
  - Concern from the individual, a knowledgeable informant, or a clinician that the person has experienced a significant decline in cognitive functioning.
  - A significant impairment in cognitive performance, as evidenced by standardized neuropsychological testing (preferred) or another quantified clinical assessment.

Additional diagnostic criteria include:

- Deficits in cognitive functioning interfere with daily activities, such as paying bills.
- Cognitive deficits do not occur exclusively with delirium and are not better explained by another mental disorder.

Specifiers include whether the disorder is due to*:

- Alzheimer's disease
- Frontotemporal degeneration
- Lewy body disease
- Vascular disease
- Traumatic brain injury
- Substance/medication use
- HIV infection
- Prion disease
- Parkinson's disease
- Huntington's disease
- Another medical condition
- Multiple etiologies
- Unspecified etiology

*The diagnostic criteria for each of these specifiers are listed separately in the DSM-5-TR and are discussed further in the proceeding sections of this desk reference guide.

Additional specifiers include:

- Without behavioral disturbance
- With behavioral disturbance
- Mild
- Moderate
- Severe

It's essential to make note of the following coding procedures*:

- Coding is based on whether the neurocognitive disorder is mild or major and on the associated etiological medical code; in the case of a neurocognitive disorder that is substance/medication-induced, coding is based on the type of substance causing the disorder.

*See the full DSM-5-TR manual for a table that provides exact codes for the neurocognitive disorders.

## Mild Neurocognitive Disorder

A mild neurocognitive disorder is diagnosed in people who show:

- Modest cognitive decline when compared to a previous level of functioning. The decline must be in one or more cognitive domains, including complex attention, executive function, learning and memory, language, perceptual-motor, or social cognition. To be diagnosed, a person must meet the following criteria:
- Concern from the individual, a knowledgeable informant, or a clinician that the person has experienced a modest decline in cognitive functioning.
- A modest impairment in cognitive performance, as evidenced by standardized neuropsychological testing (preferred) or another quantified clinical assessment.

Additional diagnostic criteria include:

- Deficits in cognitive functioning do not interfere with being independent in everyday activities, such as paying bills, but require greater effort or accommodations.
- Cognitive deficits do not occur exclusively with delirium and are not better explained by another mental disorder.

Specifiers include whether the disorder is due to*:

- Alzheimer's disease
- Frontotemporal degeneration
- Lewy body disease
- Vascular disease
- Traumatic brain injury
- Substance/medication use
- HIV infection
- Prion disease
- Parkinson's disease
- Huntington's disease
- Another medical condition
- Multiple etiologies
- Unspecified etiology

*The diagnostic criteria for each of these specifiers are listed separately in the DSM-5-TR and are discussed further in the proceeding sections of this desk reference guide.

Additional specifiers include:

- Without behavioral disturbance
- With behavioral disturbance

It's important to make note of the following coding procedures*:

- Coding is based on whether the neurocognitive disorder is mild or major and on the associated etiological medical code.

- In the case of a neurocognitive disorder that is substance/medication-induced, coding is based on the type of substance causing the disorder.

*See the full DSM-5-TR manual for a table that provides exact codes for the neurocognitive disorders.

## Major or Mild Neurocognitive Disorder Due to Alzheimer's Disease

Major or mild neurocognitive disorder due to Alzheimer's disease is diagnosed in people who meet the criteria for a major or mild neurocognitive disorder and who experience:

- An insidious onset and gradual progression of cognitive impairment.
- Impairment in at least 2 cognitive domains for a diagnosis of major neurocognitive disorder.
- To be diagnosed with major neurocognitive disorder, a person must meet the following criteria for either probably Alzheimer's disease or possible Alzheimer's disease*:
  - Probable Alzheimer's disease is diagnosed if a person has evidence of a genetic mutation that causes Alzheimer's disease based on family history or genetic testing OR they show the 3 following symptoms:
    - Clear evidence that memory and learning have declined, as well as evidence of decline in at least one other cognitive domain, as evidenced by detailed history or serial neuropsychological testing.
    - Cognitive decline is steadily progressive and gradual, without extended plateaus.
    - There is no evidence of mixed etiology; for instance, no neurodegenerative or cerebrovascular disease exists.
  - Possible Alzheimer's disease is diagnosed if neither of the following criteria above are met.

To be diagnosed with mild neurocognitive disorder, a person must meet the following criteria for probable Alzheimer's disease:

- Evidence of a genetic mutation that causes Alzheimer's disease, based on family history or genetic testing.
- Alternatively, a person may be diagnosed with possible Alzheimer's disease in the context of a mild neurocognitive disorder if:
  - There is no history of a genetic mutation that causes Alzheimer's disease and all 3 of the following criteria are met:
    - Clear evidence shows a decline in memory and learning.
    - Cognitive decline is steadily progressive and gradual, without extended plateaus.
    - There is no evidence of mixed etiology; for instance, no neurodegenerative or cerebrovascular disease exists.

The following coding notes are helpful, but one must see the full coding table in the DSM-5-TR manual for proper coding:

- Major neurocognitive disorder due to probable or possible Alzheimer's disease, with behavioral disturbance, is coded first as G30.9 for Alzheimer's disease, followed by F02.81.
- Major neurocognitive disorder due to probable or possible Alzheimer's disease, without behavioral disturbance, is coded first as G30.9 for Alzheimer's disease, followed by F02.80.
- Mild neurocognitive disorder due to Alzheimer's disease is coded as G31.84. For mild neurocognitive disorder due to Alzheimer's disease, with and without behavioral disturbance are not coded.
- Severity mild, moderate, and severe specifiers are used but cannot be coded for major neurocognitive disorder.

## Major or Mild Frontotemporal Neurocognitive Disorder

Major or mild frontotemporal neurocognitive disorder is diagnosed in people who: meet the criteria for major or mild neurocognitive disorder and who experience an insidious onset and gradual progression of symptoms.

A person must meet the criteria for either the disorder's behavioral or language variant.

- To meet the criteria for the behavioral variant, a person must show at least 3 of the following behavioral symptoms:
  - Disinhibited behavior.
  - Apathy or inertia.
  - Loss of sympathy or empathy.
  - Behavior that is perseverative, stereotyped, or compulsive/ritualistic.
  - Dietary changes and hyperorality.
- To meet the criteria for the language variant, a person must show a prominent decline in language abilities, which involves difficulties with speech production, word finding, object naming, grammar, or word comprehension.
- Another criterion for diagnosis is that a person experiences relative sparing of learning and memory and perceptual-motor function.
- Probable frontotemporal neurocognitive disorder is diagnosed if a person has evidence of a genetic mutation that can cause this disorder based on family history or genetic testing or if there is evidence of disproportionate involvement of the frontal or temporal lobe from neuroimaging.
- Possible frontotemporal neurocognitive disorder is diagnosed when no evidence of a genetic mutation and no neuroimaging has been performed.

The following coding notes are helpful, but one must see the full coding table in the DSM-5-TR manual for proper coding:

- Major neurocognitive disorder due to probable or possible frontotemporal degeneration with behavioral disturbance is coded first as G31.09, followed by F02.81.
- Major neurocognitive disorder due to probable or possible frontotemporal degeneration without behavioral disturbance is coded first as G31.09, followed by F02.80.
- Mild neurocognitive disorder due to frontotemporal degeneration is coded as G31.84. For mild neurocognitive disorder due to

frontotemporal degeneration, with and without behavioral disturbance are not coded.
- Severity mild, moderate, and severe specifiers are used but cannot be coded for major neurocognitive disorder.

## Major or Mild Neurocognitive Disorder with Lewy Bodies

Major or mild neurocognitive disorder with Lewy bodies is diagnosed in people who:

- Meet the criteria for major or mild neurocognitive disorder and who experience an insidious onset and gradual progression of symptoms.
- To be diagnosed, a person must show a combination of core and suggestive diagnostic features for either probable or possible neurocognitive disorder with Lewy bodies.
- For probable major or mild neurocognitive disorder with Lewy bodies, the person must show 2 core features or one suggestive feature with one or more core features.
- For possible major or mild neurocognitive disorder with Lewy bodies, the person only shows one core feature or one or more suggestive features.
- Core diagnostic features are as follows:
  - Fluctuation cognition with pronounced variations in attention and alertness.
  - Detailed, well-formed recurring visual hallucinations.
  - Spontaneous parkinsonism features which develop after the cognitive decline.
- Suggestive diagnostic features are as follows:
  - Meeting criteria for REM sleep behavior disorder.
  - Severe neuroleptic sensitivity.

The following coding notes are helpful, but one must see the full coding table in the DSM-5-TR manual for proper coding:

- Major neurocognitive disorder with probable or possible Lewy bodies with behavioral disturbance is coded first as G31.83, followed by F02.21.

- Major neurocognitive disorder with probable or possible Lewy bodies without behavioral disturbance is coded first as G31.83, followed by F02.80.
- Mild neurocognitive disorder with Lewy bodies is coded as G31.84. For mild neurocognitive disorder with Lewy bodies, with and without behavioral disturbance are not coded.
- Severity mild, moderate, and severe specifiers are used but cannot be coded for major neurocognitive disorder.

## Major or Mild Vascular Neurocognitive Disorder

Major or mild vascular neurocognitive disorder is diagnosed in people who:
- Meet the criteria for major or mild neurocognitive disorder
- Show clinical features indicative of vascular etiology, as evidenced by either of the following:
  - Cognitive deficits show an onset related to one or more cerebrovascular events.
  - There is evidence of a prominent cognitive decline in complex attention, including processing speed and frontal-executive function.
- To be diagnosed, there must also be evidence from history, physical examination, or neuroimaging that cerebrovascular disease is significant enough to account for the neurocognitive deficits.
- Symptoms cannot be better explained by another brain disease or systemic disorder.
- If one of the following symptoms is present, probable vascular neurocognitive disorder is diagnosed; otherwise, a person is given the diagnosis of possible vascular neurocognitive disorder:
  - Neuroimaging evidence shows significant parenchymal injury explained by cerebrovascular disease.
  - One or more documented cerebrovascular events explain the neurocognitive syndrome.
  - There are both clinical and genetic indicators of cerebrovascular disease.

- A diagnosis of possible vascular neurocognitive disorder is given if clinical criteria are present, but neuroimaging is not available, and the temporal relationship of the symptoms has not been established as related to one or more cerebrovascular events.

The following coding notes are helpful, but one must see the full coding table in the DSM-5-TR manual for proper coding:

- Major neurocognitive disorder probably or possibly due to vascular disease with behavioral disturbance is coded as F01.51.
- Major neurocognitive disorder probably or possibly due to vascular disease without behavioral disturbance is coded as F01.50.
- Mild vascular neurocognitive disorder is coded as G31.84. For mild vascular neurocognitive disorder, with and without behavioral disturbance are not coded.
- Severity mild, moderate, and severe specifiers are used but cannot be coded for major neurocognitive disorder.

## Major or Mild Neurocognitive Disorder Due to Traumatic Brain Injury

Major or mild neurocognitive disorder due to a traumatic brain injury is diagnosed in people who have a major or mild neurocognitive disorder and who have evidence of traumatic brain injury, which occurs with an impact to the head or with other mechanisms of rapid movement or displacement of the brain within the skull.

- Traumatic brain injury involves one or more of the following symptoms:
  - Loss of consciousness.
  - Posttraumatic amnesia.
  - Disorientation and confusion.
  - Neurological signs, such as neuroimaging, show injury, cortical blindness, aphasia, weakness, apraxia, loss of balance, or other sensory loss.

- To be diagnosed, a person must show symptoms of the neurocognitive disorder immediately after the occurrence of the traumatic brain injury or immediately after recovery of consciousness
- Symptoms must persist after the acute post-injury period.

The following coding notes are helpful, but one must see the full coding table in the DSM-5-TR manual for proper coding:

- Major neurocognitive disorder due to traumatic brain injury, with behavioral disturbance, is coded first as S06.2X9S followed by F02.81 for major neurocognitive disorder, due to traumatic brain injury, with behavioral disturbance.
- Major neurocognitive disorder due to traumatic brain injury, without behavioral disturbance, is coded first as S06. 2X9S, followed by F02.80 for major neurocognitive disorder due to traumatic brain injury, without behavioral disturbance.
- Mild neurocognitive disorder due to traumatic brain injury is coded as G31.84.
- For mild neurocognitive disorder due to traumatic brain injury, with and without behavioral disturbance are not coded.
- Severity mild, moderate, and severe specifiers are used but cannot be coded for major neurocognitive disorder.

## Substance/Medication-Induced Major or Mild Neurocognitive Disorder

Substance/medication-induced major or mild neurocognitive disorder is diagnosed in people who meet the criteria for major or mild neurocognitive disorder. To be diagnosed:

- Neurocognitive impairments must not occur exclusively during delirium
- Must persist beyond the typical duration of intoxication and acute withdrawal.
- In addition, the involved substance or medication must be capable of producing neurocognitive impairment
- The timing of the neurocognitive deficits must be consistent with the timing of substance or medication use and abstinence.

The following coding notes are helpful, but one must see the full coding table in the DSM-5-TR manual for proper coding:

- For mild substance use disorder comorbid with a substance-induced major neurocognitive disorder, '1' is coded in the 4th position with mild substance use disorder before substance-induced major neurocognitive disorder.
- If the substance use disorder is moderate or severe, the 4th position is coded '2' with the appropriate substance use disorder recorded before substance- induced major neurocognitive disorder.
- If there is not a substance use disorder, the 4th position is coded '9' followed by the substance-induced major neurocognitive disorder.
- For mild substance use disorder comorbid with a substance-induced mild neurocognitive disorder, '1' is coded in the 4th position with 'mild substance use disorder' before 'substance-induced mild neurocognitive disorder.
- If the substance use disorder is moderate or severe, the 4th position is coded '2' with the appropriate substance use disorder recorded before substance- induced mild neurocognitive disorder.
- If there is not a substance use disorder, the 4th position is coded '9' followed by substance-induced mild neurocognitive disorder.
- This disorder can be specified as persistent, and coding is based on the type of substance causing impairment and whether a person has a mild, moderate, severe, or no substance use disorder.

## Major or Mild Neurocognitive Disorder Due to HIV Infection

Major or mild neurocognitive disorder due to HIV infection is diagnosed in people who meet the criteria for major or mild neurocognitive disorder and who have documented infection with HIV.

The following coding notes are helpful, but one must see the full coding table in the DSM-5-TR manual for proper coding:

- Major neurocognitive disorder due to HIV infection, with behavioral disturbance, is coded first as B20 and then as F02.81.
- Major neurocognitive disorder due to HIV infection, without behavioral disturbance, is coded as B20 and then as F02.80.

- Mild neurocognitive disorder due to HIV infection is coded as F31.84. For mild neurocognitive disorder due to HIV infection, with and without behavioral disturbance are not coded.
- Severity mild, moderate, and severe specifiers are used but cannot be coded for major neurocognitive disorder.

## Major or Mild Neurocognitive Disorder Due to Prion Disease

Major or mild neurocognitive disorder due to prion disease is diagnosed in people who meet the criteria for major or mild neurocognitive disorder and who show:

- Insidious onset of symptoms, with rapid progression of impairment being typical.
- To be diagnosed, a person must show motor features of prion disease, including myoclonus or ataxia or biomarker evidence.

The following coding notes are helpful, but one must see the full coding table in the DSM-5-TR manual for proper coding:

- Major neurocognitive disorder due to prion disease, with behavioral disturbance, is first coded as A81.9 followed by F02.81.
- Major neurocognitive disorder due to prion disease, without behavioral disturbance, is first coded as A81.9 followed by F02.80.
- Mild neurocognitive disorder due to prion disease is coded as F31.84. For mild neurocognitive disorder due to prion disease, with and without behavioral disturbance are not coded.
- Severity mild, moderate, and severe specifiers are used but cannot be coded for major neurocognitive disorder.

## Major or Mild Neurocognitive Disorder Due to Parkinson's Disease

Major or mild neurocognitive disorder due to Parkinson's disease is diagnosed in people who:

- Meet the criteria for major or mild neurocognitive disorder

- Who experience cognitive disturbance alongside established Parkinson's disease. This diagnosis is characterized by insidious onset and gradual decline in impairment.

Major or mild neurocognitive disorder due to probable Parkinson's disease is diagnosed if both of the following criteria are met; if just one is met, major or mild neurocognitive disorder due to possible Parkinson's disease is diagnosed:

- There is no evidence of mixed etiology; for instance, the person does not have another neurodegenerative or cerebrovascular disease.
- Parkinson's disease preceded the onset of the neurocognitive disorder.

The following coding notes are helpful, but one must see the full coding table in the DSM-5-TR manual for proper coding:

- Major neurocognitive disorder probably or possibly due to Parkinson's disease, with behavioral disturbance is coded first as G20 for Parkinson's disease and then as F02.81.
- Major neurocognitive disorder probably or possibly due to Parkinson's disease without behavioral disturbance is coded first as F20 for Parkinson's disease and then as F02.80.
- Mild neurocognitive disorder due to Parkinson's disease is coded as G31.84. For mild neurocognitive disorder due to Parkinson's disease, with and without behavioral disturbance are not coded.
- Severity mild, moderate, and severe specifiers are used but cannot be coded for major neurocognitive disorder.

## Major or Mild Neurocognitive Disorder Due to Huntington's Disease

Major or mild neurocognitive disorder due to Huntington's disease is diagnosed in people who meet the criteria for major or mild neurocognitive disorder and who:

- Experience an insidious onset and gradual progression of disease. To be diagnosed:
- A person must show clinically established Huntington's disease or risk for the disease based on genetic testing or family history.

The following coding notes are helpful, but one must see the full coding table in the DSM-5-TR manual for proper coding:

- Major neurocognitive disorder due to Huntington's disease, with behavioral disturbance, is coded first as G10 for Huntington's disease and then as F02.81.
- Major neurocognitive disorder due to Huntington's disease, without behavioral disturbance, is coded first as G10 for Huntington's disease and then as F02.80.
- Mild neurocognitive disorder due to Huntington's disease is coded as G31.81. For mild neurocognitive disorder due to Huntington's disease, with and without behavioral disturbance are not coded.
- Severity mild, moderate, and severe specifiers are used but cannot be coded for major neurocognitive disorder.

## Major or Mild Neurocognitive Disorder Due to Another Medical Condition

Major or mild neurocognitive disorder due to another medical condition is diagnosed in people who:

- Meet the criteria for a major or mild neurocognitive disorder
- Show evidence from physical examination, history, or laboratory findings that their cognitive decline is a result of another medical condition, such as multiple sclerosis.
- To be diagnosed, a person's cognitive symptoms cannot be better explained by another mental disorder or by another neurocognitive disorder, such as major neurocognitive disorder due to Alzheimer's disease.

The following coding notes are helpful, but one must see the full coding table in the DSM-5-TR manual for proper coding:

- Major neurocognitive disorder due to another medical condition, with behavioral disturbance, is first coded according to the other medical condition and then as F02.81.
- Major neurocognitive disorder due to another medical condition, without behavioral disturbance is first coded according to the other medical condition and then F02.80.

- Mild neurocognitive disorder due to another medical condition is coded as G31.84. For mild neurocognitive disorder due to another medical condition, with and without behavioral disturbance are not coded.
- Severity mild, moderate, and severe specifiers are used but cannot be coded for major neurocognitive disorder.
- For major or mild neurocognitive disorder due to another medical condition, use additional codes to indicate significant psychiatric symptoms due to that condition.

## Major or Mild Neurocognitive Disorder Due to Multiple Etiologies

Major or mild neurocognitive disorder due to multiple etiologies is diagnosed in people:

- Meet the criteria for a major or mild neurocognitive disorder
- Who show evidence, based on history, physical examination, or laboratory findings, that the disorder results from multiple etiologies. For example, a person with this diagnosis may have neurocognitive disorder due to Alzheimer's disease and then later develop a vascular neurocognitive disorder.

The following coding notes are helpful, but one must see the full coding table in the DSM-5-TR manual for proper coding:

- Coding is based on the etiology of the neurocognitive disorder for major neurocognitive disorder. Each etiological condition is coded. The disorder is also coded according to whether it occurs with or without behavioral disturbance.
- For mild neurocognitive disorder due to multiple etiologies, the disorder is coded as G31.84. The specific etiologies are not coded for mild neurocognitive disorder. With and without behavioral disturbance are also not coded.
- Severity mild, moderate, and severe specifiers are used but cannot be coded for major neurocognitive disorder.

## Unspecified Neurocognitive Disorder

Unspecified neurocognitive disorder (R41.9) is diagnosed in people who have:

- Symptoms characteristic of a neurocognitive disorder
- Symptoms do not meet the full criteria for any of the other disorders in this chapter.
- This category is used when the precise etiology of the neurocognitive disorder cannot be determined to make an etiological attribution.

# CHAPTER 18

# Personality Disorders

The personality disorders chapter in the DSM-5-TR includes 10 distinct personality disorders. Individuals with personality disorders show lasting patterns of behavior and inner experience that deviate significantly from what is expected in one's culture. Individuals with these disorders show behaviors and thinking patterns that are pervasive and inflexible and remain stable over time. Onset is in adolescence or early adulthood.

The personality disorders are broken down into 3 clusters. Cluster A personality disorders are described as odd/eccentric; Cluster B disorders are dramatic, emotional, or erratic; and Cluster C disorders are anxious/fearful.

Cluster A disorders include paranoid, schizoid, and schizotypal personality disorders. Cluster B disorders include antisocial, borderline, histrionic, and narcissistic personality disorders. Finally, Cluster C disorders include avoidant, dependent, and obsessive-compulsive personality disorders.

The specific disorders in this chapter are described in detail below.

## General Personality Disorder

The personality disorders chapter in the DSM-5-TR begins with a description of the general criteria for personality disorders. The criteria for a personality disorder are as follows:

- Showing a lasting pattern of inner experience and behavior that is significantly different from what is expected in one's culture

- The pattern manifests in at least 2 of the following areas:
  - Cognition
  - Affectivity
  - Interpersonal functioning
  - Impulse control

Other important criteria for a general personality disorder include:

- The pattern is inflexible and pervasive, existing across a broad range of situations.
- The pattern causes significant distress or dysfunction.
- The pattern is stable and has a long duration, beginning at least by adolescence or early adulthood.
- Another mental disorder does not better explain the pattern, and it is not attributed to the effects of a substance or other medical condition.

## Paranoid Personality Disorder

Paranoid personality disorder (F60.0) is diagnosed in people who show:

- A pervasive pattern of distrust for other people, which involves suspiciousness and interpreting others' motives as malevolent.
- To be diagnosed, a person must show at least 4 of the following:
  - Suspecting that others are exploiting, harming, or deceiving them without sufficient evidence.
  - Experiencing preoccupation with unjustified doubts about the loyalty of other people.
  - Being hesitant to confide in others due to irrational fear that information shared will be used maliciously.
  - Finding demeaning or threatening meanings hidden within benign remarks or events.
  - Persistently holding grudges.
  - Perceiving that others are attacking one's character or reputation, even when this is not apparent, and quickly reacting with anger or a counterattack.
  - Experiencing recurrent suspicions that one's spouse or sexual partner is being unfaithful without evidence of such.

- Symptoms cannot occur exclusively in the context of schizophrenia, a bipolar disorder or depressive disorder with psychotic features, or another psychotic disorder.

## Schizoid Personality Disorder

Schizoid personality disorder (F60.1) is diagnosed in people who show:

- A pervasive pattern of detachment from social relationships as well as restricted emotional expression in interpersonal contexts.
- To be diagnosed, a person must show at least 4 of the following:
  - Showing a lack of desire for and enjoyment of close relationships, including family relationships.
  - Nearly always choosing solitary activities.
  - Showing little to no interest in experiencing sexual relationships with others.
  - Finding very few to no activities to be pleasurable.
  - Showing a lack of close friends or confidants aside from close family.
  - Showing indifference to praise or criticism from others.
  - Being emotionally cold, detached, or flat in affect.
- To be diagnosed with schizoid personality disorder, symptoms cannot occur exclusively in the context of schizophrenia, a bipolar disorder or depressive disorder with psychotic features, another psychotic disorder, or autism spectrum disorder.

## Schizotypal Personality Disorder

Schizotypal personality disorder (F21) is diagnosed in people who show:

- A pervasive pattern of deficits in social and interpersonal areas, involving discomfort with and reduced capacity for close relationships
- Cognitive or perceptual distortions and eccentric behavior.
- To be diagnosed, a person must show at least 5 of the following:
  - Ideas of reference.
  - Odd beliefs and magical thinking that affect behavior and are inconsistent with cultural norms.

- Experiencing unusual perceptions, including bodily illusions.
  - Odd thinking and speech, which can involve vague, circumstantial, metaphorical, overelaborate, or stereotyped speech.
  - Suspicious or paranoid thinking patterns.
  - Inappropriate or constricted emotions.
  - Odd, eccentric, or peculiar behavior or appearance.
  - Lacking close friends or confidants, aside from close family.
  - Experiencing excessive social anxiety that does not improve with familiarity and that tends to involve paranoid fears rather than negative judgments about oneself.
- Symptoms cannot occur exclusively in the context of schizophrenia, a bipolar disorder or depressive disorder with psychotic features, another psychotic disorder, or autism spectrum disorder.

## Antisocial Personality Disorder

Antisocial personality disorder (F60.2) is diagnosed in people who:

- Show a pervasive pattern of disregard for and violation of the rights of others since the age of 15.
- To be diagnosed, a person must show at least 3 of the following symptoms:
  - Failing to comply with social norms related to lawful behavior, which manifests as repeatedly performing acts that can lead to arrest.
  - Deceitfulness, which manifests as repeatedly lying, using aliases, or conning others for personal gain or pleasure.
  - Acting impulsively or failing to plan.
  - Irritable and aggressive behavior, which manifests as repeated physical fights or assaults.
  - Recklessly disregarding others' safety.
  - Showing consistent irresponsibility, which manifests as repeatedly failing to maintain work behavior or meet financial obligations.
  - Showing a lack of remorse when hurting, mistreating, or stealing from others.
- To be diagnosed with conduct disorder, a person must be at least 18 years of age

- There must be evidence that the person had conduct disorder before age 15.
- Furthermore, antisocial behavior cannot occur solely during schizophrenia or bipolar disorder.

## Borderline Personality Disorder

Borderline personality disorder (F60.3) is diagnosed in people who show:

- An ongoing pattern of instability in relationships, self-image, and affect, combined with significant impulsivity.
- To be diagnosed, a person must show at least 5 of the following symptoms:
  - Frantic efforts to prevent actual or perceived abandonment.
  - Shows a pattern of unstable, intense interpersonal relationships that involve alternating between idealizing and devaluing others.
  - Having an identity disturbance, characterized by significant, persistent instability in self-image or sense of self.
  - Showing impulsivity in at least 2 areas that can be self-damaging, such as spending, sex, substance misuse, reckless driving, or binge eating.
  - Repeated suicidal behavior, gestures, threats, or self-mutilating behavior.
  - Instability in affect characterized by significant reactivity of food; for instance, showing intense episodes of anxiety lasting a few days.
  - Feeling chronically empty.
  - Showing inappropriate, extreme anger or having trouble controlling anger.
  - Experiencing transient, stress-related paranoid ideation or severe dissociative symptoms.

## Histrionic Personality Disorder

Histrionic personality disorder (F60.4) is diagnosed in people who show:

- An ongoing pattern of excessive emotionality and attention-seeking behavior.

- To be diagnosed, a person must show at least 5 of the following symptoms:
    - Becoming uncomfortable in situations in which one is not the center of attention.
    - Interacting with others in an inappropriately seductive or provocative manner.
    - Showing rapidly shifting, shallow emotions.
    - Using physical appearance to draw attention to oneself.
    - Speaking in a way that is excessively impressionistic and lacking in detail.
    - Tendency toward self-dramatization, theatricality, and exaggerated emotional expression.
    - Being suggestible.
    - Viewing relationships as more intimate than they are.

## Narcissistic Personality Disorder

Narcissistic personality disorder (F60.81) is diagnosed in people who show:

- An ongoing pattern of grandiosity, need for admiration, and lack of empathy for others.
- To be diagnosed, a person must show at least 5 of the following symptoms:
    - Grandiose sense of self-importance.
    - Preoccupations with unlimited success, power, brilliance, beauty, or ideal love.
    - Believing one is special and unique can only be understood by and should associate with others who are exceptional and high-status.
    - Requiring excessive admiration.
    - Being entitled and expecting extremely favorable treatment or automatic compliance with one's demands.
    - Exploiting others for one's gain.
    - Showing a lack of empathy.
    - Frequently becoming envious of others or believing others are envious of them.
    - Being arrogant and haughty in attitude or behavior.

## Avoidant Personality Disorder

Avoidant personality disorder (F60.6) is diagnosed in people who show:

- An ongoing pattern of social inhibition, feelings of inadequacy, and hypersensitivity to negative evaluation.
- To be diagnosed, a person must show at least 4 of the following:
  - Avoiding work-related activities that involve significant contact with others due to fear of criticism, disapproval, or rejection.
  - Being unwilling to become involved with others unless one is confident they will be liked.
  - Restraining oneself in interpersonal relationships due to the fear of being shamed or ridiculed.
  - Preoccupation with being criticized or rejected in social situations.
  - Showing inhibition in new interpersonal situations due to feeling inadequate.
  - Viewing oneself as socially inept, unappealing, or inferior to other people.
  - Being unusually reluctant to take risks or participate in new activities due to fear of embarrassment.

## Dependent Personality Disorder

Dependent personality disorder (F60.7) is diagnosed in people who show:

- An ongoing, excessive need to be cared for, leading to submissive and clingy behavior coupled with fears of separation.
- To be diagnosed, a person must show at least 5 of the following symptoms:
  - Difficulty making everyday decisions unless one has an excessive amount of advice or reassurance from others.
  - Requiring others to assume significant responsibility for most areas of one's life.
  - Having difficulty expressing disagreement with others due to fear of losing support or approval.
  - Having difficulty starting projects or doing tasks independently due to a lack of self-confidence in judgment or abilities.

- Taking excessive steps to obtain nurturing and support from others to the extent that one volunteers to do unpleasant things.
- Experiencing feelings of being uncomfortable or helpless when alone due to extreme fear of being unable to care for oneself.
- Unrealistic preoccupation with fears of being left to care for oneself.

## Obsessive-Compulsive Personality Disorder

Obsessive-compulsive personality disorder (F60.5) is diagnosed in people who show:

- An ongoing pattern of preoccupation with orderliness, perfectionism, and interpersonal control, which interferes with flexibility, openness, and efficiency.
- To be diagnosed, a person must show at least 4 of the following symptoms:
  - Preoccupation with details, rules, lists, order, organization, or schedules to the point that one loses the central point of the activity.
  - Showing perfectionism that interferes with the completion of a task.
  - Excessive devotion to work and productivity, to the point that leisure activities and friendships suffer, without obvious economic necessity.
  - Being overly conscientious, scrupulous, or inflexible regarding morality, ethics, or values.
  - Being unable to get rid of worn out or worthless items, even when they have no sentimental value.
  - Being hesitant to delegate tasks to others or work with others unless they perform tasks precisely the way one does.
  - Spending as little money as possible and viewing money as something to be hoarded for future crises.
  - Showing rigid and stubborn behavior.

## Personality Disorder Due to Another Medical Condition

Personality disorder due to another medical condition (F07.0) is diagnosed in people with:

- A persistent personality disturbance indicative of a change from a person's previous personality pattern. In children, this manifests as a significant deviation from normal development or a substantial change in the child's typical behavioral patterns, lasting for at least a year.

Other important criteria for diagnosis include:

- Evidence based on history, physical examination, or laboratory findings that the disturbance is directly related to another medical condition.
- Another mental condition cannot better explain the disturbance.
- The disturbance does not occur only in the context of delirium.
- The disturbance causes significant distress or dysfunction.

Specifiers include:

- Labile type
- Disinhibited type
- Aggressive type
- Apathetic type
- Paranoid type
- Other type
- Combined type
- Unspecified type

Personality disorder due to another medical condition is also coded according to the other medical condition involved.

## Other Specified Personality Disorder

Other specified personality disorder (F60.89) is diagnosed in people with:

- Symptoms characteristic of a personality disorder who do not meet full diagnostic criteria for another personality disorder
- The clinician specifies the reason full diagnostic criteria are not met for another disorder.

## Unspecified Personality Disorder

Unspecified personality disorder (F60.9) is diagnosed in people with:

- Symptoms characteristic of a personality disorder who do not meet the full criteria for another personality disorder
- The clinician does not specify the reason full diagnostic criteria are not met for another disorder.

# CHAPTER 19

# Paraphilic Disorders

The paraphilic disorders chapter in the DSM-5-TR includes disorders involving intense, persistent sexual interest that deviates from typical sexual behavior between consenting human partners. Some of these disorders can consist of behavior that is harmful to others or that rises to the level of being a criminal offense.

The specific disorders in this chapter are described in detail below.

## Voyeuristic Disorder

Voyeuristic disorder (F65.3) is diagnosed in people who:

- Experience recurrent, intense sexual arousal from observing an unsuspected person who is naked, in the process of undressing, or engaged in sexual activity; this can involve fantasies, urges, or behaviors over at least 6 months.

Other important criteria for diagnosis include:

- A person has acted on their sexual urges with a non-consenting person, or the urges result in significant distress or dysfunction.
- The person involved is at least 18 years of age.

Specifiers include:

- In a controlled environment
- In full remission

## Exhibitionistic Disorder

Exhibitionistic disorder (F65.2) is diagnosed in people who:

- Over at least 6 months, experience recurrent, intense sexual arousal from exposing their genitals to an unsuspecting person; this can involve fantasies, urges, or behaviors.

Other important criteria for diagnosis include:

- A person has acted on their sexual urges with a non-consenting person, or the urges result in significant distress or dysfunction.

Specifiers include:

- Sexually aroused by exposing genitals to prepubertal children
- Sexually aroused by exposing genitals to sexually mature individuals
- Sexually aroused by exposing genitals to prepubertal children and to physically mature adults
- In a controlled environment
- In full remission

## Frotteuristic Disorder

Frotteuristic disorder (F65.81) is diagnosed in people who:

- Over at least 6 months, experience recurrent, intense sexual arousal from touching or rubbing against a non-consenting person; this can manifest as fantasies, urges, or behaviors.

Other important criteria for diagnosis include:

- A person has acted on their sexual urges with a non-consenting person, or the urges result in significant distress or dysfunction.

Specifiers include:

- In a controlled environment
- In full remission

## Sexual Masochism Disorder

Sexual masochism disorder (F65.51) is diagnosed in people who:

- Over at least 6 months, experience recurrent, intense sexual arousal from being humiliated, beaten, bound, or otherwise made to suffer; this can manifest as fantasies, urges, or behaviors.

Other important criteria for diagnosis include:

- A person has acted on their sexual urges with a non-consenting person, or the urges result in significant distress or dysfunction.

Specifiers include:

- With asphyxiophilia
- In a controlled environment
- In full remission

## Sexual Sadism Disorder

Sexual sadism disorder (F65.52) is diagnosed in people who:\

- Over at least 6 months, experience recurrent, intense sexual arousal from the physical or psychological suffering of another person; this can manifest as fantasies, urges, or behaviors.

Other important criteria for diagnosis include:

- A person has acted on their sexual urges with a non-consenting person, or the urges result in significant distress or dysfunction.

Specifiers include:

- In a controlled environment
- In full remission

## Pedophilic Disorder

Pedophilic disorder (F65.4) is diagnosed in people who:

- Over at least 6 months, experience recurrent, intense sexually arousing fantasies, sexual urges, or behaviors related to sexual activity with a prepubescent child or children, generally aged 13 or younger.

Other important criteria for diagnosis include:

- The person has acted on their sexual urges, or the urges cause significant distress or dysfunction.
- The person is at least 16 years old and at least 5 years older than the children who are the object of sexual fantasies.

Specifiers include:

- Exclusive type
- Nonexclusive type
- Sexually attracted to males
- Sexually attracted to females
- Sexually attracted to both
- Limited to incest

An individual in late adolescence involved in an ongoing sexual relationship with a 12-year-old or 13-year-old does not meet diagnostic criteria.

## Fetishistic Disorder

Fetishistic disorder (F65.0) is diagnosed in people who:

- Over at least 6 months, experience recurrent, intense sexual arousal from either the use of nonliving objects or from a highly specific focus on nongenital body parts; this can manifest as fantasies, urges, or behaviors.

Other important criteria for diagnosis include:

- The fantasies, urges, or behaviors cause significant distress or dysfunction.

- The fetish objects are not limited to pieces of clothing used in cross-dressing or to devices, such as a vibrator, used for tactile genital stimulation.

Specifiers include:
- Body part(s)
- Nonliving object(s)
- Other
- In a controlled environment
- In full remission

## Transvestic Disorder

Transvestic disorder (F65.1) is diagnosed in people who, over at least 6 months, experience recurrent, intense sexual arousal from cross-dressing; this can manifest as fantasies, urges, or behaviors.

Other important criteria for diagnosis include:
- The fantasies, urges, or behaviors cause significant distress or dysfunction.

Specifiers include:
- With fetishism
- With autogynephilia
- In a controlled environment
- In full remission

## Other Specified Paraphilic Disorder

Other specified paraphilic disorder (F65.89) is diagnosed in people who show:
- Symptoms of a paraphilic disorder
- Symptoms do not meet the full criteria for another disorder in this chapter

- The clinician specifies the reason criteria for another paraphilic disorder are not met.
- The disorder is recorded as "other specified paraphilic disorder" followed by the specific reason, which could be zoophilia, for instance.

## Unspecified Paraphilic Disorder

Unspecified paraphilic disorder (F65.9) is diagnosed in people who show:

- Symptoms characteristic of a paraphilic disorder
- Symptoms do not meet the full criteria for another disorder in this chapter
- The clinician does not specify the reason the person does not meet the criteria for another disorder in this chapter.

# CHAPTER 20

# Cultural Diagnoses and Conditions for Further Study, Interviewing

Beyond the different categories of disorders and diagnostic criteria for these various disorders, the DSM-5-TR contains additional sections, including cultural concepts of distress, conditions for further study, and tools that can be used for interviewing. This chapter will briefly discuss these areas of the DSM-5-TR. Clinicians or students can benefit from quickly referencing these areas, whether using this book as a tool in clinical practice or as a resource to guide the process of studying for an exam.

## Cultural Concepts of Distress

The DSM-5-TR contains a section on cultural concepts of distress, which describes ways that individuals in different cultures may experience and describe suffering, behavioral problems, or upsetting thoughts and emotions. Clinicians need to understand how other cultures describe these concepts of distress.

Some of the most commonly reported cultural concepts of distress include:

- **Ataque de nervios:** This syndrome is described in Latinx cultures, and it involves intense emotional upset, which may include acute anxiety, anger, or grief. Common symptoms include uncontrollable screaming and shouting, attacks of crying, trembling, a sensation of heat in the chest that rises to the head, and verbal and physical aggression. Some people experiencing these ataques may also show dissociative symptoms, fainting episodes, and suicidal behavior. While this concept of distress does not align precisely with any disorders in the DSM-5-TR, it has some overlap with panic disorder, other specified or unspecified dissociative disorder, and conversion disorder.

- **Dhat syndrome:** This syndrome originated in South Asia and describes clinical symptoms in young men who attribute their distress to semen loss. It involves diverse symptoms, such as anxiety, fatigue, weakness, weight loss, depression, erectile dysfunction, and various somatic complaints. This syndrome has some relationship to major depressive disorder, persistent depressive disorder, generalized anxiety disorder, somatic symptom disorder, illness anxiety disorder, erectile disorder, early ejaculation, and other specified or unspecified sexual dysfunction.

- **Hikikomori:** Hikikomori is a Japanese syndrome involving severe social withdrawal, which may lead to entirely ceasing in-person interaction with other people. This syndrome typically affects adolescent and young adult males, who seclude themselves in their rooms. People experiencing this syndrome tend to engage in virtual social exchanges and do not desire to attend school or work. This condition is related to social anxiety disorder, major depressive disorder, generalized anxiety disorder, posttraumatic stress disorder, autism spectrum disorder, schizoid personality disorder, avoidant personality disorder, and schizophrenia.

- **Khyal Cap:** This condition involves what are called "wind attacks," and it is found in Cambodian cultures. Symptoms are similar to those seen in panic attacks, including dizziness, palpitations, shortness of breath, cold extremities, tinnitus, and neck soreness. The focus of these attacks is on concern that a wind-like substance may rise in the body along with blood and cause issues like compression of the lungs. Related conditions in the DSM-5-TR include panic attack, panic disorder, generalized anxiety disorder, agoraphobia, posttraumatic stress disorder, and illness anxiety disorder.

- **Kufungisisa:** This syndrome occurs among the Shona of Zimbabwe, and it is believed to cause anxiety, depression, and somatic problems.

It is an idiom of psychosocial distress, and it is used to describe interpersonal and social difficulties like marital problems and unemployment. Individuals with this syndrome may ruminate on upsetting thoughts and experience anxiety symptoms, excessive worry, irritability, depressive symptoms, and posttraumatic stress symptoms. Related conditions include major depressive disorder, persistent depressive disorder, generalized anxiety disorder, posttraumatic stress disorder, obsessive-compulsive disorder, and prolonged grief disorder.

- **Maladi Dyab:** Maladi Dyab is used in Haitian communities to describe a range of medical and psychiatric disorders. According to cultural concepts, envy and malice cause people to harm enemies by using sorcerers to inflict illnesses like psychosis, depression, social or academic failure, and inability to engage in activities of daily living. People who are attractive, wealthy, intelligent, or successful are vulnerable to this condition. This condition may result in a misdiagnosis of delusional disorder or schizophrenia due to the cultural explanation that supernatural forces cause illness.
- **Nervios:** The term nervios is used in Latinx cultures to describe a general state of vulnerability to stressful life experiences and challenging life circumstances. Nervios is used to refer to a range of symptoms related to emotional distress, somatic disturbance, and difficulty with functioning. Symptoms linked to nervios include headaches, irritability, gastrointestinal disturbances, sleep problems, nervousness, tearfulness, concentration problems, trembling, tingling sensations, and dizziness. Nervios can be used to describe a range of different psychological conditions. It is related to major depressive disorder, persistent depressive disorder, generalized anxiety disorder, social anxiety disorder, other specified or unspecified dissociative disorder, somatic symptom disorder, and schizophrenia.
- **Shenjing Shuairuo:** This condition means "weakness of the nervous system" in Mandarin Chinese. This condition involves 3 out of 5 symptom clusters: weakness, emotions, excitement, nervous pain, and sleep. This condition has been used less commonly in recent years and has been replaced by idioms of depression and anxiety. Related conditions in the DSM-5-TR include major depressive disorder,

persistent depressive disorder, generalized anxiety disorder, somatic symptom disorder, social anxiety disorder, specific phobia, and posttraumatic stress disorder.

- **Susto:** Susto refers to distress and misfortune within Latinx cultures. It means "fright" and is an illness linked to a frightening event in which the soul is believed to leave the body, resulting in unhappiness and sickness. Symptoms can emerge from days to years after the frightening event occurs. Symptoms include appetite disturbances, lack of or excessive sleep, bad dreams, sadness, low self-worth, lack of motivation, interpersonal sensitivity, muscle aches and pains, cold extremities, pallor, headache, stomachache, and diarrhea. This condition is linked to major depressive disorder, posttraumatic stress disorder, other specified or unspecified trauma and stressor-related disorder, and somatic symptom disorder.
- **Taijin Kyofusho:** Taijin kyofusho is a Japanese disorder involving anxiety about and avoidance of interpersonal interactions because of concern that a person's appearance and actions are inadequate or offensive to others. The "sensitive type" of this syndrome experiences extreme social sensitivity, while the "offensive type" is worried about offending others. People experiencing this syndrome may worry about body odor, blushing, inappropriate gaze, or awkward facial expressions. The syndrome may also include features of body dysmorphic disorder and delusional disorder. Other related conditions include social anxiety disorder and obsessive-compulsive disorder.

## Conditions for Further Study

The DSM-5-TR includes a section on conditions for further study, which presents proposed criteria for conditions requiring additional research. Additional research will allow experts to understand these conditions better and make decisions about placing them in future editions of the DSM. The conditions in this section should not be used in clinical practice to make a diagnosis; they are simply proposed criteria for disorders that may or may not appear in future editions of the DSM.

The conditions in this section are detailed below.

## Attenuated Psychosis Syndrome

Attenuated psychosis syndrome is proposed for diagnosis in people who show at least one of the following:

- Attenuated delusions.
- Attenuated hallucinations.
- Attenuated disorganized speech.

Other proposed criteria for diagnosis include:

- Symptoms are present at least once a week for the previous month.
- Symptoms began or grew worse within the previous year.
- Symptoms are severe enough to be distressing and disabling and to warrant clinical attention.
- Symptoms are not better explained by another mental disorder like a depressive or bipolar disorder with psychotic features and are not the result of the physiological effects of a substance or another medical condition.
- The person has never met the criteria for another psychotic disorder.

## Depressive Episodes with Short-Duration Hypomania

This disorder is proposed for diagnosis in people who have experienced at least one major depressive episode during their lifetimes, as evidenced by meeting the following criteria:

- 5 or more of the following symptoms have been present during the same 2 weeks and represent a change in functioning, with at least one of the symptoms being either a depressed mood or loss of interest or pleasure:
  - Depressed mood most of the day, nearly every day.
    - Significantly diminished interest or pleasure in all or most activities most of the day, almost every day.
  - Significant weight loss or weight gain.
  - Insomnia or excessive sleep nearly every day.
  - Psychomotor agitation or retardation nearly every day.
  - Fatigue or loss of energy nearly every day.
    - Feeling worthless or experiencing excessive or inappropriate guilt nearly every day.

- ○ Difficulty with thinking or concentrating nearly every day.
  - ■ Recurrent thoughts of death, recurring suicidal ideation without a plan, or a suicide attempt or specific plan for completing suicide.
- In addition to the depression symptoms above, a person must experience at least 2 lifetime episodes of hypomania that meet the required hypomania symptoms but do not last long enough to meet the criteria for a hypomanic episode (i.e., lasting at least 2 days but fewer than 4 consecutive days).

To meet the criteria for hypomania, a person must show a distinct period in which mood is abnormally and persistently elevated, expansive, or irritable, and energy or activity levels are increased. To show evidence of this, a person must experience at least 3 of the following symptoms, or at least 4 if one's mood is irritable:

- Inflated self-esteem or grandiosity.
- Reduced need for sleep.
- Being more talkative than usual or feeling pressure to keep talking.
- Showing a flight of ideas or feeling that one's thoughts are racing.
- Becoming easily distracted.
- Showing an increase in goal-directed activity.
  - ○ Engaging excessively in activities that can cause painful consequences, such as unrestrained shopping sprees or foolish business investments.

## Caffeine Use Disorder

Caffeine use disorder is proposed for diagnosis in people who have a pattern of caffeine use that is problematic and leads to impairment and dysfunction; to be diagnosed, a person would need to meet at least the first 3 symptoms below, but other symptoms may also be present:

- Wanting to cut down on caffeine or making unsuccessful efforts to reduce use.
  - ○ Continuing to use caffeine, even when it causes or worsens a physical or psychological health problem.

- Experiencing withdrawal symptoms related to caffeine.
- Consuming larger amounts of caffeine than intended.
  - Continuing to use caffeine, even when it causes or worsens problems in relationships.
  - Developing a tolerance for caffeine, as evidenced by needing more to achieve the same effects or showing reduced response to the usual dose.
  - Spending a significant amount of time using or obtaining caffeine or recovering from its effects.
- Experiencing caffeine cravings.

## Internet Gaming Disorder

Internet gaming disorder is proposed for diagnosis in people who show persistent, recurrent use of Internet games, typically with other players, resulting in clinically significant impairment or distress. Diagnosis would require a person to show at least 5 of the following criteria:

- Preoccupation with Internet games, which do not involve gambling.
- Withdrawing symptoms like irritability and sadness when not gaming.
  - Developing a tolerance, so that a person must spend increased amounts of time gaming to get the same satisfaction.
- Being unable to control participation in Internet games.
  - Losing interest in previous hobbies and other sources of entertainment due to gaming.
  - Continuing to play Internet games, despite knowing it is causing psychosocial problems.
  - Being untruthful to family members, therapists, and others about the amount of gaming time.
- Using games to cope with negative moods.
  - Jeopardizing or losing relationships, jobs, or educational/career opportunities due to gaming.

The disorder is proposed to have mild, moderate, and severe specifiers.

## Neurobehavioral Disorder Associated with Prenatal Alcohol Exposure

Neurobehavioral disorder associated with prenatal alcohol exposure is proposed for diagnosis when there is evidence, based on maternal self-report, medical or other records, or clinical observation, that a person was exposed to more than minimal levels of alcohol during gestation, including before when a mother recognized she was pregnant.

Additional proposed criteria for diagnosis include:

- Impaired neurocognitive functioning, which requires evidence of at least one of the following:
  - Impaired global intellectual performance, typically manifesting as an IW of 70 or below.
  - Impaired executive functioning which can manifest through symptoms like poor planning and organizational skills and difficulties with self-control).
  - Impaired ability to learn, as evidenced by a learning disability or lower academic achievement than expected.
  - Impaired memory.
  - Impaired visual-spatial reasoning, which can manifest as being unable to differentiate left from right or as disorganized drawings.
- Impairment in self-regulation, as evidenced by at least one of the following:
  - Impaired ability to regulate mood or behavior.
  - Deficit in attention.
  - Impaired impulse control.
- Impaired adaptive functioning, as evidenced by at least 2 of the following:
  - Communication deficits, such as delayed language acquisition.
  - Impaired social communication and interaction.
  - Impaired daily living skills, such as delaying toilet training.
  - Impaired motor skills, such as poor fine motor development or deficits in coordination and balance.

To be diagnosed, the onset of symptoms would have to occur during childhood.

## Suicidal Behavior Disorder

Suicidal behavior disorder is proposed for diagnosis in people who have made a suicide attempt within the previous 24 months; to be considered a suicide attempt, the actions a person took must have been with the intent to cause their death.

Other proposed diagnostic criteria include:

- The action taken does not meet the criteria for non-suicidal self-injury; this means the action was not intended to induce relief from a negative mood state or to produce a positive mood state.
- Diagnosis cannot be applied to suicidal ideation or preparatory acts.
- The attempt cannot be made during a state of delirium or confusion.
- The attempt is not completed solely for political or religious reasons.

Specifiers would include current (not more than 12 months since the last attempt) and early remission (12-24 months since the previous attempt).

## Non-Suicidal Self-Injury Disorder

Non-suicidal self-injury disorder is proposed for diagnosis in people who, over the previous year, have engaged in intentional self-injury to the surface of their body on 5 or more days; to meet criteria, the injury to the surface of the body must be of the type to likely induce bleeding, bruising, or pain. Examples include cutting, burning, stabbing, hitting, or excessively rubbing the body while expecting that the behaviors will lead to minor or moderate physical harm. With this disorder, there is no suicidal intent behind the behavior.

Other proposed diagnostic criteria include:

- The person engages in the behavior to relieve a negative feeling or cognitive state, resolve an interpersonal problem, or induce a positive feeling state.
- The intentional self-injury is linked to interpersonal problems or negative feelings or thoughts like depression, anger, tension, anxiety, self-criticism, or generalized distress experienced immediately before

the act; the person is preoccupied with the self-injury and unable to control the preoccupation before engaging in the act; or, the person frequently thinks about self-injury even when not acting on it.

To be diagnosed, the behavior would need to cause clinically significant distress or dysfunction, and it could not be socially sanctioned, meaning it could not occur in the context of body piercing or getting a tattoo. Nail biting or picking scabs do not align with this disorder's criteria.

Finally, diagnosis would not occur in people who show this behavior only during psychotic episodes, delirium, substance intoxication, or substance withdrawal. Diagnosis also would not occur in those for whom another mental disorder or medical condition better explains the behavior. This disorder also would not be diagnosed in people with neurodevelopmental disorder who show the behavior simply as part of repetitive stereotypies.

## Interviewing

Clinicians may desire tools to assist with interviewing clients to make a diagnosis. While there are specific tools that vary based on the diagnostic category (i.e., depression, anxiety disorders, personality disorders), the Level 1 cross-cutting symptom measures are a broad interview tool helpful in identifying symptoms that occur across a variety of disorders.

This tool is intended to be self-report or informant-report, meaning clinicians can interview the client directly to complete the tool or ask an informant, such as a parent or spouse, to provide answers. Each item on the Measure is ranked on a 5-point scale, with 0 meaning the symptom is experienced none or not at all and 4 meaning severe or experienced nearly every day. If a person scores a 2 or higher on any item within a domain, in most cases, this suggests the need for further evaluation. In the case of substance use, a score of 1 or higher indicates a need for further evaluation. For example, if a person ranks themselves as 2 or higher on the question, "How often during the past 2 weeks have you been bothered by having little interest or pleasure in doing things?" this would suggest the clinician should complete a diagnostic interview for depression.

## Chapter 20: Cultural Diagnoses and Conditions for Further Study, Interviewing

For adults, the Level Cross-cutting symptom measures contain the following items, which are each scored on a scale of 0 to 4, based on how often over the previous 2 weeks a person has been bothered by each of the following:

Domain 1:

- Little interest or pleasure in doing things.
- Feeling down, depressed, or hopeless.

Domain 2:

- Feeling more irritated, cranky, or angry than usual.

Domain 3:

- Sleeping less than usual but still having a lot of energy.
- Starting lots more projects than usual or doing more risky things than usual.

Domain 4:

- Feeling nervous, anxious, frightened, worried, or on edge.
- Feeling panicked or frightened.
- Avoiding situations that make one anxious.

Domain 5:

- Unexplained aches and pains (e.g., head, back, joints, abdomen, legs).
- Feeling that one's illnesses are not being taken seriously enough.

Domain 6:

- Thoughts of actually hurting oneself.

Domain 7:

- Hearing things other people couldn't hear, such as voices, even when no one was around.
- Feeling that someone could hear one's thoughts or that one could hear what another person was thinking.

Domain 8:

- Problems with sleep that affect sleep quality overall.

Domain 9:

- Problems with memory (e.g., learning new information) or location (i.e., finding the way home).

Domain 10:

- Unpleasant thoughts, urges, or images that repeatedly enter the mind.
- Feeling driven to perform certain behaviors or mental acts over and over again.

Domain 11:

- Feeling detached or distant from oneself, one's body, one's physical surroundings, or one's memory.

Domain 12:

- Not knowing who one is or what one wants out of life.
- Not feeling close to other people or enjoying relationships with them.

Domain 13:

- Drinking at least 4 drinks of any kind of alcohol in a single day.
- Smoking any cigarettes, a cigar, or pipe, or using snuff or chewing tobacco.
- Using any of the following medicines, on one's own, without a doctor's prescription, in greater amounts or longer than prescribed: painkillers like hydrocodone or oxycodone, stimulants like methylphenidate or amphetamine/dextroamphetamine; sedatives or tranquilizers like diazepam; or drugs like marijuana, cocaine, or crack, club drugs like ecstasy, hallucinogens like LSD, heroin, inhalants or solvents like glue, or 'speed,' such as crystal meth.

For adults, the specific domains are as follows:

- 1- Depression
- 2-Anger
- 3-Mania
- 4-Anxiety
- 5-Somatic symptoms
- 6-Suicidal ideation
- 7-Psychosis
- 8-Sleep problems
- 9-Memory
- 10-Repetitive thoughts and behaviors
- 11-Dissociation
- 12-Personality functioning
- 13-Substance use

For children, the Level 1 Cross-cutting symptom measures contain the following items, which are each scored on a scale of 0 to 4, based on how often over the previous 2 weeks a person has noticed their child doing each of the following:

Domain 1:

- Complaining of stomachaches, headaches, or other aches and pains.
- Saying he or she is worried about their health or about getting sick.

Domain 2:

- Having problems sleeping—trouble falling asleep, staying asleep, or waking up too early.

Domain 3:

- Having problems paying attention when they are in class, doing their homework, reading a book or playing a game.

Domain 4:

- Having less fun doing things than they used to.
- Seeming sad or depressed for several hours.

Domain 5/6:

- Seeming more irritated or easily annoyed than usual.
- Seeming angry or losing their temper.

Domain 7:

- Starting lots more projects than usual or doing more risky things than usual.
- Sleeping less than usual for them but still having lots of energy.

Domain 8:

- Saying they felt nervous, anxious, or scared.
- Not being able to stop worrying.
- Saying they couldn't do things they wanted to or should have done because they may feel nervous.

Domain 9:

- Saying that they heard voices- when there was no one there- speaking about them, telling them what to do, or saying bad things to them.
- Saying that they had a vision where they were completely awake- that is, they saw something or someone that no one else could see.

Domain 10:

- Saying that they had thoughts that kept coming into their mind, such as that they would do something terrible or that something bad would happen to them or someone else.
- Saying that they felt the need to check on certain things over and over again, like whether a door was locked or whether the stove was turned off.
- Seeming to worry a lot about things they touched being dirty, germs, or poisoned.
- Saying that they had to do things in a certain way, like counting or saying special things out loud, to keep something terrible from happening.

Domain 11:

- Having an alcoholic beverage.
- Smoking a cigarette, cigar, or pipe, or using snuff or chewing tobacco.
- Using drugs like marijuana, cocaine, or crack, club drugs like ecstasy, hallucinogens like LSD and heroin, inhalants or solvents like glue, or methamphetamine, like speed.
- Using any medicine without a doctor's prescription; for example, opioid painkillers like hydrocodone or oxycodone; stimulants like methylphenidate or amphetamine/dextroamphetamine; sedatives such as benzodiazepines (diazepam, lorazepam) and sleeping pills; or steroids.

Domain 12:

- Talking about wanting to kill themself or about wanting to commit suicide.
- EVER trying to kill themself.

For children, the specific domains are as follows:
- 1- Somatic symptoms
- 2-Sleep problems
- 3-Inattention
- 4-Depression
- 5-Anger
- 6-Irritability
- 7-Mania
- 8-Anxiety
- 9-Psychosis
- 10-Repetitive thoughts and behaviors
- 11-Substance use
- 12-Suicidal ideation/suicide attempts

# References

*American Psychiatric Association. (2022).* **Diagnostic and statistical manual of mental disorders** *(5th ed., text rev.).* https://doi.org/10.1176/appi.books.9780890425787

# CHECK OUT OUR OTHER BOOKS

**Pharmacology Review:**
A Comprehensive Reference Guide for Medical, Nursing, and Paramedic Students

**Pharmacology Review:**
A Comprehensive Reference Guide for Medical, Nursing, and Paramedic Students: Workbook

**Medical Terminology:**
The Best and Most Effective Way to Memorize, Pronounce and Understand Medical Terms (2nd Edition)

**Medical Terminology:**
The Best and Most Effective Way to Memorize, Pronounce and Understand Medical Terms: Workbook

Scan the QR Code

**EKG/ECG Interpretation:**
Everything you Need to Know about the
12 - Lead ECG/EKG Interpretation and
How to Diagnose and Treat Arrhythmias
(2nd Edition)

**EKG/ECG Interpretation:**
Everything you Need to Know about the
12 - Lead ECG/EKG Interpretation and
How to Diagnose and Treat Arrhythmias:
Workbook

**Advanced Cardiovascular Life Support**
Provider Manual - A Comprehensive
Guide Covering the Latest Guidelines

**Advanced Cardiovascular Life Support:**
Provider Manual - A Comprehensive
Guide Covering the Latest Guidelines:
Workbook

**Lab Values:**
Everything You Need to Know
about Laboratory Medicine and its
Importance in the Diagnosis of Diseases

**Medical Surgical Nursing:**
Test your Knowledge with Comprehensive Exercises in Medical-Surgical Nursing: Workbook

**Basic Life Support:**
Provider Manual - A Comprehensive Guide Covering the Latest Guidelines

**Anatomy & Physiology**
The Best and Most Effective Way to Learn the Anatomy and Physiology of the Human Body: Workbook

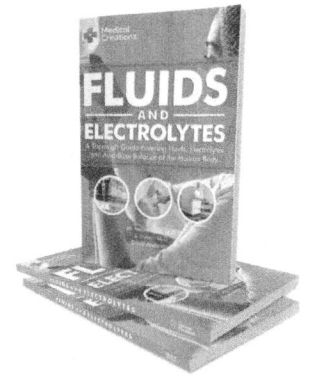

**Fluids and Electrolytes:**
A Torough Guide covering Fluids, Electrolytes and Acid-Base Balance of the Human Body

**Suture like a Surgeon:**
A Doctor's Guide to Surgical Knots and Suturing Techniques used in the Departments of Surgery, Emergency Medicine, and Family Medicine

Made in United States
Orlando, FL
12 April 2025